IT'S ALL *LIGHT*

The Morphic Resonance of Light

– A Unified Theory

BY GLORIA PREMA

Order this book online at www.trafford.com
or email orders@trafford.com

Most Trafford titles are also available at major online book retailers.

© Copyright 2010 Gloria Prema.
All rights reserved. No part of this publication may be reproduced, stored in a retrieval system, or transmitted, in any form or by any means, electronic, mechanical, photocopying, recording, or otherwise, without the written prior permission of the author.

Printed in Victoria, BC, Canada.

ISBN: 978-1-4251-9247-1

Our mission is to efficiently provide the world's finest, most comprehensive book publishing service, enabling every author to experience success. To find out how to publish your book, your way, and have it available worldwide, visit us online at www.trafford.com/10510

Trafford rev. 2/1/2010

 www.trafford.com

North America & international
toll-free: 1 888 232 4444 (USA & Canada)
phone: 250 383 6864 ♦ fax: 812 355 4082

DEDICATION

To my father, Edward Pratt, who was my first source of inspiration, and to my children, Richard, Julia and Edward, who have always been my beacons of light.

Contents

DEDICATION	V
ACKNOWLEDGEMENTS	XI
INTRODUCTION	XIII
CHAPTER 1 - THE BEGINNING - IN A NUTSHELL	1
CHAPTER 2 - THE MISUNDERSTANDINGS	5
CHAPTER 3 - THE PHYSICS	21
CHAPTER 4 - THE DISHARMONY	37
CHAPTER 5 - HOLISTIC CONSCIOUSNESS	43
CHAPTER 6 - A LITTLE HISTORY LESSON	52
CHAPTER 7 - IT'S ALL RELATIVE	56
CHAPTER 8 - A DIALECTIC THEORY	60
CHAPTER 9 - THE SOUL IS A HOLOGRAM	63
CHAPTER 10 - VORTEX RESONANCE	67
CHAPTER 11 - CONSCIOUSNESS IS THE KEY	71
CHAPTER 12 - THE BLASPHEMY, OR THE BEHAVIOUR OF LIGHT	80

CHAPTER 13 - MORPHIC RESONANCE OF LIGHT	91
CHAPTER 14 - THE LAW OF RESONANCE	112
CHAPTER 15 - PROOF OF SPEEDS FASTER THAN LIGHT	120
CHAPTER 16 - ELECTROMAGNETISM	129
CHAPTER 17 - GRAVITY AND THE STRONG AND WEAK NUCLEAR FORCES	137
CHAPTER 18 - DARK MATTER AND BLACK HOLES	147
CHAPTER 19 - SEEING WITH THE HEART	152
CHAPTER 20 - THE GOLDEN RATIO AND FRACTALS	156
CHAPTER 21 - REINCARNATION OR REBIRTH	164
CHAPTER 22 - THE HEART AND QUANTUM CONCEPTS	174
CHAPTER 23 - AURAS AND RADIATION	180
CHAPTER 24 - THE CAUSE OF CANCER	185
CHAPTER 25 - MUSIC OF THE SPHERES	204
CHAPTER 26 - ALL YOU NEED IS LOVE	210
REFERENCES:	213
BIBLIOGRAPHY, RECOMMENDED READING AND WEBSITES	223

Figures

FIGURE 1 - DIAGRAM OF THE ATOM	22

FIGURE 2 – THE ELECTROMAGNETIC SPECTRUM	24

FIGURE 3 – ELECTRON/PROTON BECOMING A NEUTRON/NEUTRINO	32

FIGURE 4 - TRANSVERSE WAVES OF ELECTROMAGNETISM	81

FIGURE 5 - MORPHIC RESONANCE PATTERNS CREATED BY ASCENDING FREQUENCIES	88

FIGURE 6 - THE PLATONIC SOLIDS	96

FIGURE 7 - GEOMETRICAL SHAPES CREATED BY MORPHIC RESONANCE	97

FIGURE 8 - MORPHIC RESONANCE FIELD DISTORTED BY HEAT	123

FIGURE 9 - THE MANDELBROT SET	159

FIGURE 10 - PI AND PHI (EXAMPLES OF A GOLDEN RATIO SPIRAL AND LEAF SHAPE

AND A LOGARITHMIC SPIRAL PRODUCED
BY A GOLDEN RATIO RECTANGLE) 161

FIGURE 11 – THE CHAKRA SYSTEM OF THE
HUMAN BODY 182

FIGURE 12 – KIRLIAN PHOTOS OF THE
FINGERTIPS OF A PERSON WITHOUT
CANCER AND A PERSON WITH CANCER 191

ACKNOWLEDGEMENTS

I would like to thank the great scientists, philosophers and mystics of the past whose work continues to inspire me and to those of today who are bravely bringing forward a new understanding of this cosmos which we inhabit.

I would also like to thank the staff of the Physics lab at Aberdeen University, who very kindly allowed me to use their equipment to conduct my own cymatic experiments, and with particular thanks to Bob Mowat, Physics Lab Technician, for his expert technical assistance.

Thank you too to Michael Glickman, Architect extraordinaire, for his delightful discussion on the golden ratio.

Grateful thanks also to the staff at Trafford Publishing for their friendly, helpful advice and invaluable assistance.

And finally, grateful thanks to any unseen help I may have received.

IT'S ALL *LIGHT*

INTRODUCTION

> *"If we do discover a complete theory, it should in time be understandable in broad principle by everyone, not just a few scientists. Then we should all, philosophers, scientists and just ordinary people, be able to take part in the discussion of the question of why it is that we and the universe exist. If we find the answer to that, it would be the ultimate triumph of human reason – for then we would know the mind of God."*
>
> Stephen Hawking, Physicist

This book presents a unified theory in the broadest sense, reconciling not only science, religion and spiritual philosophy, but within science, the reconciliation of the opposing theories of relativity and quantum physics and a unification of the four main forces in physics. I am calling this theory, The Morphic Resonance of Light. It explains that science and religion are polarised views of a much broader concept of universality – what may be termed spirituality, and goes on to explain the physics of spiritual phenomena, such as inspiration, prayer, healing, near-death experiences amongst others; spiritual philosophy such as karma and rebirth and the physics of achieving a state of enlightenment or realisation. In essence it is a return to the study of physis, which is where the word 'physics' comes from. Physis was the term coined by the early Greek philosophers of 6th century BC where science, religion and philosophy were all considered as one interconnected subject. Physis was the understanding of the fundamental essence of all life and that this essence was the unifying principle of things animate and inanimate, seen and unseen. It is a great pity that this understanding was lost over the centuries as humanity split into various camps and disciplines, resulting in the chasm between science and religion that we have today. Even so, great thinkers like Newton and Einstein understood the physis principle and believed in a spiritual force which permeated all things. History hails them as great scientists but they were also mystics. That term just means that they had a deeper understanding beyond the confines of a scientific discipline. Without physis there is a tendency for religion to distance itself from science and there is a tendency for science to distance itself from both religion and spirituality, often making the common mistake of confusing one with the other.

The comparison between religion and spirituality can be defined as follows: whereas religions are based on doctrines and belief systems to spread the teachings of an individual and tend to be divisive and exclusive, for example, by claiming to be the only true path; spirituality recognises there can be many paths leading to the same source which are equally valid so long as it's the right one for the individual, but crucially, it recognises the interconnectedness of all things, in other words it is beyond divisiveness and this includes embracing scientific truths as part of it. This is because a Creator or First Source would necessarily have created the scientific laws as well as all the other laws. Spirituality is about perceiving beyond duality consciousness, science versus religion for example. The theory presented in this book attempts to create a dialectic way of viewing the *seeming* opposites. The term dialectic was coined by the ancient Greeks as an explanation of using reason and discussion to discover the truth. The German philosopher, Hegel, used the term to describe a process of thought where apparent contradictions (thesis and antithesis) are seen to be part of a higher truth (synthesis) and this is the meaning applied here. The scientific laws still stand but with extensions to include unseen dimensions, or spiritual dimensions if you like, plus the addition of one new law, that of the Morphic Resonance of Light. The theory could be seen as an alternative explanation of string theory which is one of the newest theories in physics which says that at the smallest level of existence everything is formed of 'strings' which vibrate and that they exist in at least 11 dimensions. String theory, however, has no consistent logic running through it and it cannot say, for example, what the strings are made of, how they form sub-atomic particles, how form is created, how the other dimensions are accessed or what the Big Bang is. It is also in danger of becoming very complicated, as has happened with quark theory. Morphic Resonance of Light on the other hand, is a consistent, logical and more complete theory and all of the aforementioned questions, and more, can be answered.

It is explained that everything is essentially frequencies of light, spiralling wavelengths of light, travelling in harmonic multiples of the known speed of light, which become form due to sound, or resonance (continuous sounding) and that the light has consciousness. This can explain spiritual phenomena, the evolution of consciousness and the process of enlightenment, leading to realisation, one-ness, nirvana, heaven (there are many names for the same thing). In science it explains biological mysteries such as the growth of form, in chemistry it explains chemical interactions and in physics it explains the four fundamental forces in physics of gravity, electro-magnetism and the strong and weak nuclear forces, as well as giving a verbal explanation of quantum physics (which describes the very small) and reconciling this with its opposing theory of relativity (which describes the very large), to show that they are really part of the same theory, or opposite ends of the same dynamic system. At present they are irreconcilable because, although they are well tested and accepted theories of how the universe works, they seem to be polar opposites. (For example, quantum theory says we have free will because the observer affects the experiment, relativity says we do not have free will because the future is already determined). This seeming polarisation is due to a limitation of consciousness which sees things in duality. It is explained that as consciousness expands due to faster speeds of light frequencies then a dialectic understanding becomes possible. Dialectic means to see the opposites from a higher, whole perspective which is the source of the opposites and in this way the opposites can be reconciled because they can be seen to be complementary pairs.

Parts may be viewed as heretical, but heretic simply means 'those who make the choice to be an independent thinker'. It should not be viewed as an attack on the current scientific model. It is simply an attempt to expand current scientific understanding. Scientific understanding, like everything else, is constantly evolving, or at least should be. But the problem which has arisen in science today is that the current laws are held as sacrosanct and when anomalous experiences do not appear to fit those laws, then they are usually dismissed, rather than be investigated. This is not science. A similar thing happens with religions. Both camps have been incredibly dogmatic and closed-minded, generally speaking, and this has resulted in the polarisation which humanity has created for itself, science vs. religion, the 'them' and 'us' scenario.

The evidence presented in this book explains that scientific understanding must be allowed to evolve beyond the purely physical, and religious understanding must be allowed to evolve to embrace scientific truths. It is not intended to take away faith, only blind faith and replace it with knowledgeable faith. The word faith after all simply means to have 'complete trust or confidence in something or someone' according to The New Oxford Dictionary. So both religious people and scientists have faith. Knowledgeable faith and evolved scientific understanding have a point of merging. At this point of merging a state of coherence is reached. Coherence means *logical, consistent, united as, or forming a whole*, and in physics, as *having a constant phase relationship* (of waves), in other words 'united'. So with coherence a greater Awareness or Realisation of the whole is reached. This is the dialectic view, or the spiritual view. The word spiritual can carry certain associations for some people, as can the word Creator, as indicated below. Unfortunately, this is one of the problems of language but as is stated in the opening paragraph the word spiritual in the sense used here means the holistic, universal or dialectic understanding of the whole. We need to broaden our concept as to what connotes 'spiritual' or perhaps even come up with a different word altogether. As will be explained later, matter is a precipitation of unseen dimensions, spiritual dimensions if you like. We describe matter and spirit as another duality concept, but essentially they are the same thing – all Light. So perhaps we could come to the point where we can drop the words matter and spirit and simply refer to them as Light - seen or unseen Light. The word Light certainly has less emotive associations than the words matter and spirit.

In science, Isaac Newton gave the world the laws of motion applicable to low speeds. Albert Einstein later progressed these laws to explain motion at high speeds (The Special Theory of Relativity) but we must not rest on our laurels and think the laws are now complete so we should be able to explain everything with the existing laws, and if we can't explain it with the existing laws, such as the speed of light or the conservation of energy, then it can't be real and should be disregarded. We must be open to inspiration and observation to expand the current laws to explain facts of nature. This, after all, is what science is supposed to be about, but today we have a science that is often funded by commercial operations and so its driving force is profit. This detracts from the true nature of inquiry, which has to come from selflessness and it is a threat to inspiration which comes from pure observation. The present scenario has resulted in some scientists who have become mere technicians to their paymasters and others who, because their findings do not fit mainstream understanding, are forced to work behind the scenes, often without funding, given no platform to speak and often denied publishing in peer reviewed journals. This is particularly true in the fields of zero point energy and complementary medicine ...more of this later.

In this book it is hoped that, as well as opening minds and hearts, the myth that only the intellectual elite and those who wear the label of scientist can understand how the universe works, can be dispelled. I have endeavoured to make it easy reading because I believe the basic truths are simple and should be understandable by most people. Viktor Schauberger's quote at the beginning of chapter one says it all.

Nigel Calder in his book "Einstein's Universe" states "The successor to Newton and Einstein must instead look for extreme conditions which prove the equations inadequate and probe into the half-hidden electronics of the universe, short-circuiting some of its basic components. And once again it may seem that the universe quakes, as it did to Einstein's contemporaries." I don't claim to be their successor but in this book I will show that the most famous equation of all, $E=mc^2$ is inadequate to explain the anomalies in science, but if altered slightly, *can* explain them and can even explain dark matter. That mysterious invisible stuff that science tells us makes up approximately 95% of the universe. I have also probed into the 'half-hidden electronics of the universe', (in the sub-atomic particles) and by 'short-circuiting some of its basic components' (the dispensing with of quarks and the weak nuclear force in relation to the formation of neutrinos), the whole quantum theory becomes an integral part of the fractal nature of the universe and explains the expansion of the universe.

In support of my theory I would also like to invoke the Law of Economy, also known as Ockham's razor. It acquired this name after the 14th century philosopher William of Ockham, who employed the principle regularly in dealing with radical ideas, or razor sharp ideas which cut to the heart of the matter (hence the name). He felt that in order to get to the point of the matter it was necessary to accept the thesis with the minimum assumptions, or in other words, the simplest explanation was probably the correct one. The theory presented in this book has one single assumption; that lightwaves exist in harmonic multiples of the speed of light. A strong case is made for this as the whole theory is based on fractals, as well as the fact that science has already proved that faster than light speed is possible. The morphic resonance fields are fractal because harmonic sound is fractal. Harmonic and fractal means *fitting together, one fits into the next size up, ad infinitum*, like Russian dolls – the same form but different sizes fitting into each other. In mathematics a fractal is a *geometrical figure which has the same statistical character as the whole. In other words, the same pattern recurs at progressively smaller or larger scales.* And in music, for example, a harmony is *simultaneously sounding chords and chord progressions or an overtone accompanying a fundamental tone at fixed intervals,* so the chords and tones fit together in a phase relationship of waves, which is a description of coherence. So to make the whole theory consistent, the speed of lightwaves themselves would be fractal or harmonic. The speed of light fits into twice the speed of light, which fits into four times the speed of light, which fits into eight times the speed of light; harmonic octaves. Just like on a piano, each ascending note 'C' in each octave is double the frequency of the previous lower note 'C', so we have 1, then x2, x4, x8, x16, x32 etc.

The creation of matter is due to the spiralling lightwaves becoming particles due to resonance/sound. This is explained in the chapter on Morphic Resonance of Light, as sound creates form or shape. A spiralling lightwave stopped onto one point (due to sound) would naturally spin and this creates the spinning sub-atomic particles, each of which is a tiny vortex. One of the strongest cases for the existence of morphic resonance is the example of the double-slit experiment. This experiment is taught to all undergraduate physics students to show that light is both a wave and a particle, but what the experiment *really* shows is the existence of **morphic resonance as a template for physical form** as well as showing that **light has consciousness.** Morphic refers to form and resonance is sound, or more correctly re-sounding or continuing to sound. The theory is based on all forms, visible and invisible, being caused by sound, an all-pervasive resonance which gives everything in existence its frequency signature. The theory demonstrates that vortex motion (spinning) is the fundamental building method of matter, (using spinning lightwaves) and morphic resonance is the mold or template. Vortex motion follows the fractal principle, 'as above, so below' because the vortex is demonstrated in the smallest (sub-atomic particles) right up to the largest (spiralling galaxies). So both the vortex and fractal are universal principles and are inherent in resonance. Gravity and electro-magnetism are the effect of vortex motion. The strong nuclear force and the weak nuclear force can both be explained by morphic resonance The unifying of these four forces, gravity, electro-magnetism, the strong and weak nuclear force, having been the holy grail of science for some time.

However, in recent years another energy has been discovered which is counteracting the force of gravity in the universe and some scientists are calling it a fifth force, or funny energy, or negative pressure. This is what I call the spiralling light waves beyond the speed of physical light. I explain that it exists in harmonic multiples of the speed of light. It is what has formerly been called the 'ether' and these days tends to be called 'the zero-point field'. It is space itself but it is still Light and it is in a state of expansion. Contained within this space is the so-called 'dark matter' of physics. It makes up approximately 95-99% of the universe but it is 'dark' or 'invisible' and is a mystery to science. Since a particle's energy is relative to its frequency and it is argued that there are frequencies beyond the known electromagnetic spectrum because there is no actual limit to the spectrum, just a limit to which we can measure, then this allows the existence of particles beyond the known electromagnetic spectrum, creating the fine matter, the invisible 'dark matter'. It is simply finer matter existing at faster speeds of light creating the other dimensions. Anything travelling at extreme speed appears to disappear so at frequencies beyond the known electromagnetic spectrum it would be invisible, which is why it is 'dark' . The speed of light based on the known electromagnetic spectrum is a subset of a greater spectrum, so this dimension we inhabit is a subset of other dimensions and may even be a superset of other dimensions which are operating slower than ours. The point being made is, it is one of the harmonic octaves on the scale, which we happen to call the third dimension, or fourth if you include time. So this third/fourth dimension is based on the speed of light at 186,000 miles per second, the next dimension would be based on twice the speed of light, the next would be based on four times the speed of light, etc. (Remember that each octave doubles the frequency). This is one example of harmonic octaves though it could be that the dimensions follow a different set of harmonics. Ultimately, however, only two forces are needed to describe the universe – harmonic light and resonance.

The harmonic light includes all visible light and invisible light, space, the ether or the zero-point field, all energy in a state of potentiality.

The resonance includes all matter, visible and invisible, (including dark matter), following the fractal principle and the vortex principle which creates gravity, electro-magnetism and the strong and weak nuclear forces. Visible and invisible matter is spinning light, spinning due to sound, sound itself being slow light. Or, in other words,

IT'S ALL LIGHT.

I have also deliberately not included mathematics here, apart from one simple equation, for several reasons - a) it is a *verbal* unified theory – I am not a mathematician b) a mathematical presentation would exclude many people from understanding and most importantly, c) I believe physics has been unnecessarily over-complicated due to scientists using maths as the first step.

I would suggest that the understanding, the theory, has to come first. Once the theory is presented then the mathematical formulas can be worked out to follow the theory. If the theory is sound then the maths will work out.

This book came about through the realisation that higher dimensional realities can be explained from the study of quantum physics. What is happening in the microscopic world shows us that **a) light has consciousness and that it is creative b) slow light produces sound which creates the perpetual motion of particles and c) matter and light are essentially the same thing, with everything having its own unique light frequency signature.** Everything can be explained from these principles.

This theory will no doubt have to be finely tuned, altered and expanded upon in years to come as understanding increases but, hopefully, it can help pave the way to expand the current scientific laws to embrace the unseen realities, which is the home of the 'dark matter' of physics and of spiritual phenomena, both of which cannot be left out of a unified theory. Otherwise it is not a unified theory! As humans we have much evolving to do. We still have to evolve into our full brain capacity, and evolve as holistic beings, using the heart's intelligence as much as the head's. Therefore, I would ask those of you who find some of the concepts heretical or offensive to your dearly held beliefs not to nitpick, but to try to get the feel of the ideas.

<u>I would also stress that the use of the words Creator or Source are not to be taken in the traditional religious sense. Rather the use of the words are meant to convey the omniscient consciousness, the Light itself, which is inherent in everything and of which everything is a part; the ultimate source of all things visible and invisible and recognising the obvious intelligence inherent in Nature and the cosmos. It is somewhat risky to use the words at all due to the many divisive associations and the religious connotations they carry but this is one of the problems of language. Just try to remember, it's all Light.</u>

CHAPTER 1

THE BEGINNING - IN A NUTSHELL

"The majority believes that everything hard to comprehend must be very profound. This is incorrect. What is hard to understand is what is immature, unclear and often false. The highest wisdom is simple and passes through the brain directly into the heart."
Viktor Schauberger, Scientist, Inventor and Forester, (1885 - 1958)

In the book of Genesis it begins by saying 'In the beginning was the Word, and the Word was with God, and the Word was God'. The word AUM or OM in ancient scriptures is said to be the primal, eternal sound of creation. Perhaps this is 'the Word'. So in the beginning this would be the plan, the blueprint of creation, the 'Big Bang' in physics. Big Bang and the Word are both sounds. All creation emits a sound, this is its unique frequency, and it is now recognised by science that there is an all-permeating background sound in space with even black holes emitting their own unique sound, at 57 octaves below Middle C. Our auditory speech could be said to be a less-refined sound of the OM, or a lower frequency, but there are harmonics within all sounds and this can be demonstrated with overtone chanting where a certain tone produces harmonic frequencies, sounds within sounds. OM, which will be used in this text to represent that primal, cosmic sound, exists in all things at all times, visible and non-visible. All structure is held together by sound in its many resonances, from the lowest evolutionary aspect to the highest, all the way back to the Source, the OM itself. When Source breathes out OM this sound creates the subatomic particles, which are tiny vortices of energy, which then arrange and structure themselves in accordance to the Law of Resonance inherent in the OM, to form the substance of matter. The tiny vortices of energy are themselves composed of light waves spiralling throughout cosmos. **So light itself is the breath of Source. It becomes matter due to the resonance of the sound created by the breath.** Light and sound are essentially the same thing. Sound waves travel slower than light waves but if any sound is played at many octaves higher it produces coloured light which means that **sound is slow light, light is fast sound.**

The unseen light waves are currently known as space, the seen light waves are what we know as light, both waves and particles. Light is a spiralling wave but becomes a particle due to resonance (continuous sounding) so particles of matter are spinning lightwaves. The resonance creates the particular structure out of the spinning particles because sound/resonance creates geometric forms. The spin of the particles is related to the frequency. This means the particle is both spinning on its axis and vibrating up and down at the same time at a certain number of times per second (this is the frequency), thus a particle has both spin and frequency, or rotation and vibration, and there is a band of frequencies relative to the third dimension. This is the electromagnetic spectrum. The frequencies of the third dimension are relative to the speed of light at 300,000 kilometres per second or 186,000 miles per second. Higher dimensions vibrate at frequencies relative to harmonic multiples of the speed of light. This is because harmony means, as quoted in the Oxford dictionary – 'the combination of simultaneously sounded musical notes to produce chords and chord progressions having a pleasing effect and the quality of forming a pleasing and consistent whole'. As stated above, everything is formed from light and sound. **Both are harmonic – progressions but all simultaneously existing**, like the concept of Russian dolls. One size fits into the next and so on, but all are contained within the whole and all exhibit self-similarity. Self-similarity is inherent in fractals (the smallest is the same shape as the largest). Perhaps this is an explanation of being 'made in the image of the Creator'. In octaves of sound, the first note of a new octave is double the frequency of the previous octave so this is the same as saying that the next higher dimension is based on a multiple of the speed of light, say twice the speed of light, the following would then be four times the speed of light, then eight times the speed of light, etc. If you can imagine, all the waves would be able to fit into, or embed within each other due to the phase relationship of the waves.

With the creation of form, when the resonance remains the same the form stays the same. When the resonance changes the form changes or decays and the particles radiate away back into a spiralling wave. Energy becomes matter and matter becomes energy, $E=mc^2$. All energy in the universe is dynamic and has the appearance of being conserved within the physical dimension. It exists in every dimension and there are many dimensions, and dynamically interacts with every dimension including the physical, of which we are most familiar. Until quite recently it was quite heretical in science to speak of other dimensions but this has now become accepted language due to the work in quantum mechanics, in particular quantum computers and string theory.

Zero point energy or free energy or over-unity energy, all names for the same thing, can be tapped from space or the higher dimensions to be used in the physical dimension to power technology. Healers also tap into these higher dimensions. Further explanations of these will be given later in the text. In the following theory the equation $E = mc^2$ evolves into $E =$

mc^n (n representing harmonic multiples) to account for the zero point or higher dimensional energy which is also the source of the invisible matter or so-called 'missing' dark matter which is creating the mysterious extra gravity in the universe. This finer invisible matter is in existence in the faster frequencies beyond the known electromagnetic spectrum. The 'n' in the above equation becomes a cosmological variable. In other words, variable speeds of light which accounts for the time differentials of relativity which states that past, present and future all co-exist.

The equation further evolves into $C = mc^n$ where energy is shown to be equivalent to consciousness. Consciousness is shown to be equivalent to light, for example, in quantum physics, light particles (photons) make decisions and since the prime requirement for decision-making is consciousness then light and consciousness must be equivalent. Consciousness 'C' equals matter 'm' times harmonic speeds of light 'c^n' - It's All Light.

The OM contains within it the whole plan of creation; the blueprint, it is the ultimate harmony. All forms and structures resonate to their part of the Plan. This pertains not only to the physical dimension, but also to the higher (or faster) dimensions all the way back to Source. At Source, Reality Is Omniscient Consciousness, but containing within it the potentiality for all expressions, in order for Life to Experience, Return to Source with Evolutionary Gains, and hence Expand Creation throughout Cosmos.

CHAPTER 2

THE MISUNDERSTANDINGS

> *"Thoughts that have important consequences are always simple. All my thinking could be summed up with these words: 'Since corrupt people unite amongst themselves to constitute a force, then honest people must do the same.' It is as simple as that."*
> Count Leonid Tolstoy, Philosopher and Author (1828 – 1910)

It is important to begin by saying that as science presently stands there are many laws and statements made about observations from experiments, speaking particularly about physics, but often no explanation as to why or how these factors occur. To simply make a statement such as 'light is both a wave and a particle', or 'gravity waves are extremely weak' or 'matter is frozen energy', or 'gravity can arrest light to create a black hole' does not explain why these things happen. They are not explanations, just statements, and some are questionable at that. For example, it is not known if there even *are* such things as gravity waves. To make statements like these does not allow a person to understand how the universe works. And this lack of understanding is often at the very foundation of dogma and doctrine. This creates a danger of rigid thinking and intolerant attitudes. A similar phenomenon occurs in religion where adherents can often repeat scriptures verbatim but struggle to apply them to daily life or even fully understand what they mean. This is 'blind faith' but as will be explained, some scientists are just as guilty of 'blind' faith. The word 'faith' after all simply means to have complete trust or confidence in someone or something. Religious scriptures and scientific statements are both unhelpful unless they can be explained. A statement is not necessarily an explanation.

There are also some scientific laws which are contradictory, or at least partial. Such as Newton's law of gravity, which states that all things must fall at the same rate. This law is based entirely on non-spinning objects. But we know that, for example, a spinning top is heavier than a non-spinning top which means that the spinning affects gravity. Einstein's theory of gravity states that a rotating or spinning mass has a different effect on space and time from one which is not spinning. The Penguin Dictionary of Science states that gravity is reduced to a very small extent by the centrifugal force caused by the Earth's rotation (for an object at rest on its surface). It also says that in order to stay in orbit round the Earth, or other planet, an orbiting body would have to achieve a velocity which will produce a centrifugal force which *exactly balances the force of gravity*. A force which exactly balances the force of gravity, is levity. This is the same as saying that complementary spinning creates levity. So we can see that spinning and complementary spinning are fundamental to the understanding of gravity. Yet these concepts are not generally taught, probably because not enough is known about gravity and little or nothing about levity, well not in mainstream education at any rate. Some physicists are using terms like 'gravity waves' and 'gravitons' when talking of gravity, which are pure conjecture. The point is, it is a very incomplete understanding which is taught as 'the laws of gravity'

Another questionable scientific 'fact' is the non-existence of an ether or aether, a subtle energy existing in space or a 'fifth force' as Aristotle named it. An experiment to detect if there was such a thing as ether, was carried out in 1886. The ether, or subtle energy was a popular idea in science in the 18th and 19th centuries and was an accepted theory by such eminent scientists as James Clerk Maxwell (1831-1879), the Scottish physicist, who developed the idea of a *field*, and went on to explain electromagnetic phenomena using the field theory and the field theory later became an essential basis of Albert Einstein's (1879-1955) work. Like Einstein, Sir Isaac Newton (1642 – 1727) also believed in the idea of an ether, or underlying spiritual force, as discussed in The Introduction, which he believed filled all space. The ether experiment referred to above was carried out by Albert Michelson (1852-1931) and Edward Morley (1838-1923) who measured the speed of light in the direction of the Earth's rotation and against the direction of rotation. They detected no difference in the speed and concluded that there was no ether. This one single experiment is now taken as conclusive evidence against the ether theory, despite the fact that one of the most basic tenets of science is that experiments must be replicable and that only when they have been replicated does it become a scientific fact. This experiment was based on the assumption that the ether, if there was one, was static. They did not account for the fact that the ether might be in motion. So, in effect, it was only half an experiment.

Recent decades have produced two more experiments which have, in fact, detected ether. One was carried out in 1986 by E.W. Silvertooth from the United States, and reported in Nature magazine that year. Using more sophisticated equipment than Michelson and Morley, he did detect what he called an ether drift in relation to the Earth's motion at a speed of 378 kilometres per second. The second experiment was conducted a year later in 1987 by S. Marinov of Austria who confirmed the finding but with a slightly different speed of 386 kilometres per second. The important point is that they did detect ether. Nevertheless, science institutions *still* teach that there is no ether and many mainstream scientists become emotionally hysterical at the mere mention of the word. It seems these findings have been ignored, probably because the word 'ether' is, in their minds, associated with spiritualism. Spiritualism, which embraces other dimensions or unseen realities, was a popular idea in the 17th - 19th centuries, even by scientists such as Isaac Newton and Johannes Kepler (1571-1630). But what I find curious is that mainstream scientists now talk about a mysterious energy in space which seems to be causing the universe to expand at an ever-increasing acceleration. An energy seems to be counteracting the attracting force of gravity and they don't know what it is. This news was presented to the world in 1998 by astrophysicist Saul Perlmutter and his team from the Lawrence Berkeley National Laboratory. It has been called a negative pressure, a vacuum energy, even 'funny energy' which sounds very scientific! Anything will do but, heaven forbid, not that word 'ether'.

We know that space is not a vacuum. We know that space is the precursor to matter. We know that matter consists primarily of space. We know that matter is in constant vibrational motion. It is no radical assumption to suggest that space is also in motion. The energy fields affecting matter exist in space. Albert Einstein, I believe, understood this concept as he said "we may, therefore, regard matter as being constituted by the regions of space in which the field is extremely intense…There is no place in this new kind of physics both for the field and matter, for the field is the only reality." Another way of saying it, coined by the physicist David Ash is, 'matter is dense space, space is sparse matter'. And as I will explain in this book, 'the field' which Einstein referred to, is light. The zero point field referred to in Chapter 1 is light. The ether is light. We know that light has no rest energy – it is in constant motion. We know through quantum physics that the zero point field is in constant motion. We know that sub-atomic particles are in constant motion and science doesn't know how or why or where it comes from. But even though this motion is perpetual there is still a tendency in science to say that perpetual motion doesn't exist and some free thinking inventors who have invented devices which work on perpetual motion have had patents refused on the basis that perpetual motion is impossible! Light has no rest energy – it *is* perpetual motion. Since visible light, the ether or zero-point field and sub-atomic particles are all in constant or perpetual motion, is this not a very strong argument that they *are* all Light? In Hindu philosophy the great cosmic cycles of expansion and contraction are the in breath and the out breath of the cosmic creator. Likewise in this theory, the breath of Source is the light which is creating the expansion of the universe. Perhaps at some time a contraction will occur, to be followed by another expansion, and so on. This idea is consistent with everything else in nature which follows cycles and rhythms.

This book presents a theory that states that all of space and matter is in constant motion. The ether and zero point field are just other names for the subtle, non-physical, spiralling waves of light, the fifth dimension and beyond. This is where the 'dark matter' in physics is found. So it is important to understand that our scientific laws need to evolve beyond the physical. The laws work to a certain level but do not explain so much of life experiences (this is why there are still many unsolved mysteries in science). Mainstream scientists at present are trying to explain the mysteries using the existing physical laws and introducing more and more complexities into those laws, but it will never work. They will not find the answers until they embrace the fact that most of reality is non-physical. We like to think that science is at the forefront of embracing new ideas and discovering new things but unfortunately this is not always so. Scientists can be just as entrenched in their dogma as anyone else and can be as emotionally hysterical as anyone else when their particular worldview is threatened. There can be intense resistance to new ideas and several examples are given later. On the surface it can appear that scientists are coming up with new ideas and new discoveries but if you look closely you will see that oftentimes they are still very fixed within the old paradigm ideas, trying to make new discoveries fit into the known laws by introducing more and more complexity. The quark theory is a classic example of this. There is a very strong resistance in mainstream science to examine new evidence which may put into question existing understanding of the known laws, even though much of this new evidence has been collected by highly qualified scientists. Orthodoxy, it seems, is the biggest challenge to new ideas. Fortunately however, not all scientists follow orthodoxy and some real breakthroughs are being made, for example, with zero point energy technology

Zero point technology requires the acknowledgement of non-physical or higher-dimensional energy. These are often called over-unity devices or perpetual motion devices because they seem to produce energy from nowhere, in other words more energy comes out than goes in, or appears to, and in science this goes against the Conservation of Energy Law. But perpetual motion is already an established fact, in the form of sub-atomic particles, as already discussed. The electrons, protons and neutrons of the atom are in constant motion. Where is that energy coming from? They are in perpetual motion because they are made of light, vibrating to sound, which is itself part of the light, or slow light. So this is another contradiction in science which says that perpetual motion devices cannot exist because perpetual motion is impossible and yet at the same time it is known that it does exist in the sub-atomic world. The reality of the zero point energy/higher dimensions has been suppressed for decades by mainstream science and by governments. The reasons why are complex, not least of which is that, as already noted, science has a history of resisting new ideas. In his book *The Structure of Scientific Revolutions*, author Thomas Kuhn outlines how unscientific scientists can become when their own world view is threatened.

It's necessary to give here a brief description of why people feel so threatened. First of all the subject of professional jealousy arises. It's a common trait of human nature to want recognition and validation but jealousy is a step beyond this, when a person resents someone else having something because they don't have it themselves and want to take it away from them or undermine them in some way in order to feel better about themselves, or in order not to feel overlooked. This, of course, is not exclusive to scientists and happens in day-to-day living. Most of us will have experienced it. But when it happens amongst scientists it can hold back new discoveries.

The other threat comes from new or different ideas from those that a person may have invested a lifetime in and this is due to the self-image we build up about ourselves. Take, for example, a scientist. She/he may have built their career based on knowledge acquired. They have an emotional attachment to their knowledge, career, qualifications, status, position, prestige, money (linked to the status or position) and even their personal life is linked in to all of it. They have an image of themselves as belonging to particular groups; a scientific discipline, etc. or if it is a male scientist he may have an image that science is the domain of men and that women have no place in it. Unfortunately, sexism is still alive and well in some quarters, so this type of man would feel very threatened by a woman who makes an original discovery. It wouldn't fit with his self-image. Or say, if someone were to present evidence which transcends known scientific laws, this would rock the boat for the scientists with strong attachments to their acquired knowledge and they would likely reject the evidence saying that it can't be true because it breaks scientific laws and they would be unlikely to even try to replicate the experiment, already having decided it was impossible. This most unscientific behaviour is often what passes as science these days! What's really happening is their world view is threatened. It's like sticking a pin in a balloon to their way of thinking because they have so much invested in their particular view. This becomes their conditioning. And this is the same threat felt by the religious person; someone with a strong attachment to their particular religion and its doctrines, who sees any kind of change or differing viewpoint as threatening. Just a little thought experiment here …. all religions are awaiting the return of their saviour. What if that saviour was already here and they didn't recognise the fact because of their religion's interpretation of scripture, the interpretation having been passed on by tradition, which, if they adhere to that tradition, stops them thinking for themselves or observing for themselves, so they don't actually see what is before them. We tend to think and see what we are conditioned to think and see.

Charles Darwin recorded in his diaries that when he sailed into Patagonia, the islanders there could not see his ship 'The Beagle' as it sailed toward the shore because ships were unknown to them and they had no reference to go by. The ship could only be seen by the local shaman. A 'shaman' is 'a seer'. When someone has rid themselves of much of their conditioning and sees a clearer reality, they are known as a 'shaman' or a 'seer' depending on what part of the world you come from. This is a natural part of an evolving consciousness. The shaman on this island was able to describe different parts of the ship, comparing the parts to objects that they were familiar with. Eventually, over several days, the islanders were able to 'see' the ship emerging into sight. The shaman had taught them to 'see' the ship. So, unless a person undertakes steps to free themselves of conditioning (some 'how tos' are described later), then it is likely that they are stuck in a place which is determined by their knowledge of the past, which is what tradition is. The word tradition comes from the Latin 'tradere' meaning 'betrayal of the present'. Interesting ! So, for the people awaiting the return of their saviour, if it doesn't happen according to the way they have been told by their elders, whose knowledge is based on interpretations of scriptures passed down by tradition, perhaps they will fail to recognise the saviour when he/she appears.

It is not generally realised that knowledge, which attachment and tradition are based on, is a thing of the past. Knowledge is in the past. It is useful and necessary, particularly in regard to physical things; how to use a machine or drive a car for example, but psychologically it acts as a deterrent to awareness. It doesn't describe the present which is continuously moving and evolving. Awareness is having true perception. Perception comes from the Latin 'percipere' meaning to 'entirely understand', so awareness is to entirely understand the present. Perception or awareness is when you can be detached from your self-image and are open to new or different thoughts, new or different experiences, new or different ideas and the consciousness becomes aware of something more and this is what allows knowledge to evolve. This is rather like the practice of Zen, which means to see things as they actually are and not come to them with already formed judgements and conclusions. This is worth repeating – to see things as they actually are and not come to them with already formed judgements and conclusions. An example of Zen would be to look at a tree and, immediately the mind wants to classify it; naming it, saying whether it's deciduous or coniferous etc. but this is not 'seeing' the tree. To 'see' it you would see a dynamic, living, breathing entity of energy frequencies. That's not to say that naming, classifying it etc. is wrong, just that it is based on knowledge which is in the past. It's not 'seeing' it in the true sense, which is an ever-changing, living, breathing energy. It's rather like when older family members continually see you as the young, inexperienced teenager that you used to be and still treat you the same way. The problem is caused by them being stuck with an old image of you, so they constantly meet you with the past. They never update their image. You can become a brain scientist or an astronaut but those family members will always see you and regard you as 'wee Johnny' the awkward teenager. In their subconscious image you are always that gangly teenager. This is a classic example of being met with the past and it is a common experience. Another example of being met with the past is sexism, which is entirely due to conditioning, which comes from (faulty) knowledge of the past, passed on by tradition. So in everyday life we are constantly being met with the past, in almost every interaction, because few people have truly unburdened themselves of conditioning. Awareness, therefore, is only partial in most instances.

Having perception or awareness does not mean ignoring acquired knowledge, which would be foolish, but it is acknowledging that knowledge is incomplete and ever-changing, that there is always more to learn and to not be attached to knowledge acquired, including the laws of science as they presently stand, or religious doctrine as it presently stands. The more open-minded person will know this and therefore have more awareness. The open-minded scientists are the ones who are truly making leaps in science. So, knowledge of the past (however true or false) can be reflected in a person's religion, their profession, their political allegiance, nationalist allegiance, status in the community, etc. with present perception or awareness making a very minor impact. This may be an acceptable way of living to many, but the danger comes when Science as an institution or Religion as an institution conducts itself this way because they have such an iconic status in society.

So, to get back to my point about the suppression of the reality of zero-point energy; it is to do with threat to prevailing world views of orthodox science which is prejudiced towards a physical universe. All orthodox scientific laws are based on physical reality only, the three-dimensional world, but the phenomena of zero-point technology, which is a reality, breaks the Conservation of Energy Law as it stands and this is what gets orthodox scientists very upset. The technology is producing more energy than is going in. That energy, often called free energy, has to be coming from outwith the physical dimension and this is where the threat comes in. But perhaps the biggest resistance, mainly by governments, is due to the impact it would have on the world's economy. Free energy cannot be sold or monopolised, however the technology to use it could be. There is also the issue of 'national security'. With the mindset of humanity at present it would be developed for military purposes and I have no doubt that it is already being worked on, in secret, for that very purpose, and probably has been since the 1940s when America began to take an interest in the work of Viktor Schauberger (1885-1958) and Nikola Tesla (1856-

1943). Russia has also had the knowledge for at least that long. Both countries have developed flying discs based on their work and this would account for some of the UFOs seen over certain areas, particularly the Nevada test site in the U.S. Other UFOs are another story, but still with relevance to anti-gravity and zero-point technologies. Perhaps the subject of another book ...

Schauberger and Tesla's work was based on the vortex model which is the principle of implosion, centripetal rather than centrifugal force. Tesla invented technology based on energy flow through a coil, which is in essence, a spiral. He discovered that when energy travels in this pattern it creates enormous energy potential. If it travels centrifugally, this creates outward flowing energy, or explosion. If, on the other hand, it travels centripetally, this creates implosion or inward flowing energy, inwards towards the centre, which gains access to other dimensional timeframes (more on this later). This is consistent with Einstein's law of gravity which states that a spinning mass has a different effect on space and time from one which is not spinning. Viktor Schauberger also discovered this energy potential while employed as a forester in his native Austria. It was by observation of the natural surroundings that he discovered this principle of the vortex spiralling motion of water. He initially discovered it by watching trout in a stream. They appeared to be able to remain motionless against a fast flowing current and even to travel against it without effort. He discovered that the water was flowing through their gills in a corkscrew manner, creating vortexial energy flows from their gills and that this was what created the effortless propulsion of the trout against the current. He built models of this principle in his laboratory but was never able to control the energy flow produced. His models would often take off and crash through the roof.

After the second world war American scientists heard of his work and enticed him across to America. They have been working on this ever since but keeping a lid on their findings. Viktor Schauberger had discovered limitless, clean energy propulsion but he became increasingly worried by the direction that science was heading and was deeply perturbed for many years by what he could see was the increasing degradation of the environment and felt that science and technology were directly responsible. He was one of the first outspoken voices on environmental destruction and wrote extensively on the importance of maintaining healthy waterways. He maintained that natural watercourses must not be interfered with as water is a natural medium for frequency carrying which means it is an information carrier and that it must be allowed to flow in its natural state to maintain vibrancy (optimal frequency for health), its natural state of flow being that of a spiralling motion. Observe any waterway from the air and it displays a snake-like meandering pattern. Or observe steam (or more correctly water vapour) rising from a hot bath - as it rises it swirls around in a spiralling fashion. This is because water always follows the path of least resistance, which is a spiral. He was correct that water is an information carrier as this is the principle of homeopathy and this will be discussed further in the chapter on Morphic Resonance of Light. And when you consider that our bodies are composed of approximately 85 % water, the link to health is obvious. It became clear to Viktor Schauberger and his son Walter, who accompanied him on his trip to the States, that his work may not be carried on by the most ethically driven scientists, and he returned to his native Austria dejected and downhearted and died shortly afterwards, some say of a broken heart. What he and Nikola Tesla had discovered was a way to tap into the higher dimensional energy, or the zero point field and water is only one of the carriers of this energy. As described in the introduction, everything has a frequency signature and water is able to absorb frequencies which it comes into contact with.

There are many scientists who have been working with the zero point field for a long time and many books and papers have been published. A sample of these are noted [1 – 15]. Machines have been built which operate on zero point energy and these are well documented. One example is in Brian O'Leary's books. A former NASA astronaut and a scientist of the highest integrity, in his books 'The Second Coming of Science', 'Exploring Inner and Outer Space' and 'Miracle in the Void', he outlines his global travels where he witnessed these devices operating. Some inventors have had their equipment confiscated and labs closed down, or threatened into silence, others have had some support from their governments, such as in India and Japan. Brian O'Leary has bravely championed this cause for years to try to bring forward this knowledge and has often met with intense resistance from the scientific community, many of whom are quick to debunk something they have little or no knowledge of. As he said "perhaps the most common error created by the debunkers is the erroneous assumption that if these machines were real, they would have heard about it." The reason people haven't heard of it is because of the suppression. Those that *have* heard of it but try to debunk it are suffering from the self-image threat, discussed earlier. However, as the maxim goes "there is nothing so powerful as an idea whose time has come". Zero point knowledge and applications are unstoppable. It can be tapped either by technology or consciousness, the latter is not new. The technology, on the other hand, will revolutionise life on Earth and it looks like Japan may be at the forefront of producing the first marketable products.

However, because of the danger of military applications and the revolution of the economy it will have to be handled carefully. We, as a species, will have to change. Any advanced species looking down on us would probably conclude that we are too immature to responsibly handle this technology without making some changes. Look at our track record. The past century has been one of the bloodiest in history, yet we call ourselves 'civilised'. Inequalities between rich and poor continue to rise with 30,000 children dying *every single day* from hunger ….. on a planet that is bulging with food, yet we call ourselves 'democratic'. The environment is heavily polluted, with 13 million tons of toxic chemicals released into it *every day*, unreplaceable resources are used up with only token commitments made to sustainable resource use, yet we call ourselves 'intelligent'.

I am sometimes ashamed to be a member of the human race, yet we have enormous potential within us to make changes for the better and I believe it's already happening and growing in momentum. Like the zero point technology, it is unstoppable because it is evolution. The timing is now for minds and hearts to open. It is nothing short of a scientific and spiritual revolution. With the implementation of zero point technology (which is non-polluting) we will have a clean environment, we will have sustainable resource use, we will have healthier lives and we will leave behind us, for future generations, a healthy planet. Perhaps one of the biggest mistakes people make is to equate all technology with progress. Technology does not necessarily equal progress and some technology is very harmful. But on the other hand if technology is used with sound, compassionate values it can bring great benefits to humanity. We must always question new technology as to the intention behind it. Some new technologies have the sole motive of profit as the driving force and this is the wrong use of science. But scientists, like adherents to religion, cover a broad spectrum from open to closed, in mind and heart. At the open end there are already some geniuses out there, taking science to new heights. Using not only the intellect of the head, but also the intellect of the heart to understand and explain this universe and our part in it. As will be explained, the heart and the energy of love are crucial to our next leap in consciousness, with the resultant evolvement of scientific understanding. Logic alone is not enough to carry us forward. It is vacuous without feeling experience, perhaps best described as compassion, and I believe that Einstein had some understanding of this. He said "propositions arrived at by purely logical means are completely empty of reality. It is very difficult to explain this feeling to anyone who is entirely without it. I maintain that cosmic religious feeling is the strongest and noblest incitement to scientific research."

Einstein believed in a God who was omnipresent, even in the very fabric of matter and that human minds were part of God's mind. This echoed the writings of the Jewish philosopher, Baruch Spinoza, whose work he admired. The philosophy of the 'open' scientists mentioned above reflect this They are beginning to understand omniscience and omnipresence and humanity's place in all of it. They are scientists who have a more holistic understanding. Many have had spiritual experiences which their scientific training cannot begin to address and this has led them to, not turn their backs on science, but to expand scientific understanding, as they have realised there is much more to reality than previously acknowledged. As Einstein noted, "it is very difficult to explain this feeling to anyone who is entirely without it" and so it is usually not taken seriously by those who have not had the experience. This has resulted in many having to turn their backs on mainstream institutions with the resultant loss of funding and peer ridicule. Or they have just been quietly ignored. Lynne McTaggart in her book 'The Field' has done much to highlight the work of some of these scientists. They have done extensive work and study of consciousness and the zero point field. Some are beginning to see their work be taken seriously. An international science conference was held in December 2001 to look into advanced space propulsion where the zero point field was one of the factors considered to achieve this. NASA was in attendance and one of the things they have been investigating is the creation of wormholes in space as a means of deep space propulsion.

Wormholes are a scientific theory popularised in the novel 'Contact' by Carl Sagan where the heroine travels at faster than light speed to another dimension. This was also made into a film by the same name featuring Jodie Foster as the heroine. The idea was first proposed by the physicist John Wheeler of Princeton University. The idea being that between atoms there are bubbles of space and on occasion if two bubbles join, a sort of tunnel would appear and because science believes space is curved (folding over like a piece of paper) it should be possible to wormhole through to a far distant point in the galaxy, ignoring any time restrictions. Work is ongoing to try to find a way to stretch the tunnel wide enough for a human being to enter and pass through to the other side because the wormhole is only the size of an atom. There are two problems with this, however. First, wormholes have never been seen, they are simply a theory. It is not known if there *are* bubbles between atoms. Secondly, the scientists working on this are still 'thinking within the box' as it were, still trying to solve this problem with third dimensional physics.

The theory I present in this book says that space is curved due to the morphic resonance fields and the inherent spiralling nature of light. The explanation of this is given later in the text. Because all morphic fields originate from one source (the Big Bang) they must all be interconnected, or embedded in some way, or all part of the one pattern, the overall harmony. This, therefore, could enable access anywhere, so this point at least could support the wormhole idea. However, my theory goes on to describe that consciousness operating at faster speeds is fundamental to this access. At present, as already discussed, science is thinking only in third dimensional terms, trying to figure out a way to physically stretch something the size of an atom to the size of a human being. My assertion is that it will not be a purely physical, third dimensional process. However, whatever the process comes to be known as, it is not a question of if, but when, we will have faster than light speed travelling. The Warp speeds of Star Trek may not be fantasy for too much longer, or teleportation for that matter. Both will be possible when faster than light speeds and non-locality are fully understood. (Non-locality is described in the next chapter). The first teleportation experiments have already been conducted at the Australian National University (ANU). After early experiments in 1998 at the California Institute of Technology showed that it should be possible to teleport a laser beam, some 40 laboratories around the world have been trying to achieve this. In 2002 the ANU managed to make a beam of light disappear then reappear at a different location [16].

CHAPTER 3

THE PHYSICS

"We bring into existence what we think. For as he thinketh in his heart, so is he."
The Bible, Proverbs 23:7

Werner Heisenberg (Heisenberg's Uncertainty Principle) and Nobel prize winner for physics, said "at the sub-atomic level there is no such thing as an exact science. Such was the power of our consciousness that we couldn't observe anything at that level without changing it. What we observe is not nature itself, but nature exposed to our method of questioning. Therefore, quantum physics leads us to the only place there is to go – ourselves."

The Atom

Matter, as we know, is made of atoms, the atoms themselves are made of even smaller particles, sub-atomic particles, called neutrons, protons and electrons. The neutrons and protons together make the nucleus of the atom with the electrons orbiting around the nucleus, like a mini solar system.

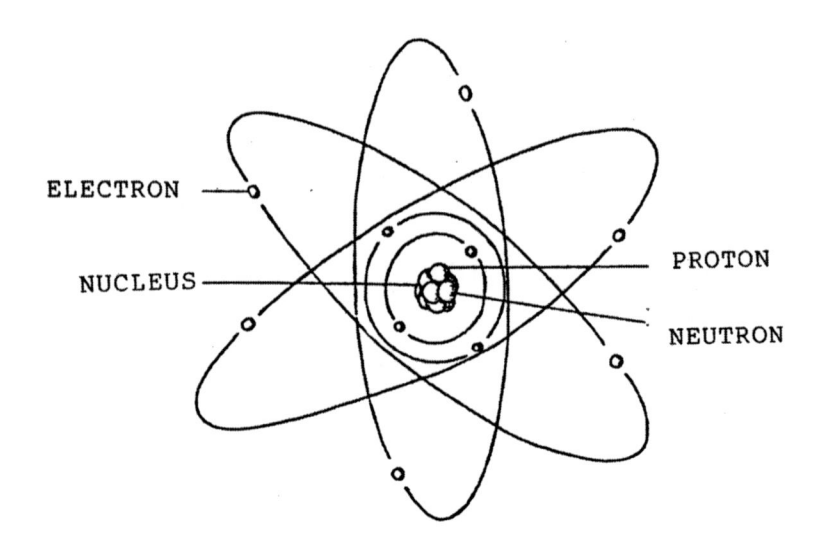

FIGURE 1 - DIAGRAM OF THE ATOM

The electrons orbit in 'shells' at various distances from the nucleus, rather like planets orbit at different distances from the sun and each 'shell' contains a fixed amount of electrons. The electrons, however, are not static, they can join with the protons in the nucleus to make a neutron or they can be annihilated if they meet an anti-electron, which is given the name positron, meaning an electron with a positive charge. Electrons are negatively charged. Electrons also give off photons (particles) of light, which can then be absorbed by other electrons or other particles. An atom itself is neutral; it has no charge. This is because it contains an exact balance of positive and negative particles. However, as has been stated, it is not a static thing; with exchange of photons, annihilation and electrical attraction going on, we can see that atoms are dynamic systems of force. As well as this going on *inside* the atom, they also vibrate to a particular frequency. In other words, they jump up and down and the number of times they do this per second is the frequency. Ordinarily, when a particle or molecule is caused to increase its frequency, this creates a temperature increase and this is what happens in cooking. So increased frequency means increased temperature and for cells this creates entropy, which means increasing degradation. This is why the human body must remain within a very narrow range of temperature, verging around 37.5° C, otherwise ill health results.

The photoelectric effect

The energy, or spin, of particles is linked to their frequency (in physics this is known as the photoelectric effect). This means particles have both rotation (spin) and vibration (frequency). So the spin and frequency of particles have to be in relative synchronisation to prevent entropy (decomposure) or incoherence (disorder). There is a band of frequencies relative to the third dimension, from the short-wavelength of gamma rays, which have wavelengths below 10^{-14} m, up to the long-wavelengths of radio waves, which have wavelengths greater than about 300 m. This range is known as the electromagnetic spectrum. All of these types of radiation travel at the same speed in a vacuum, which is the speed of light at 300,000 km or 186,000 miles per second but they differ in their frequencies, which is the number of waves per second that they achieve. As you can see from the diagram in Figure 2, only a tiny sliver of this spectrum is visible to us, giving us our colour spectrum.

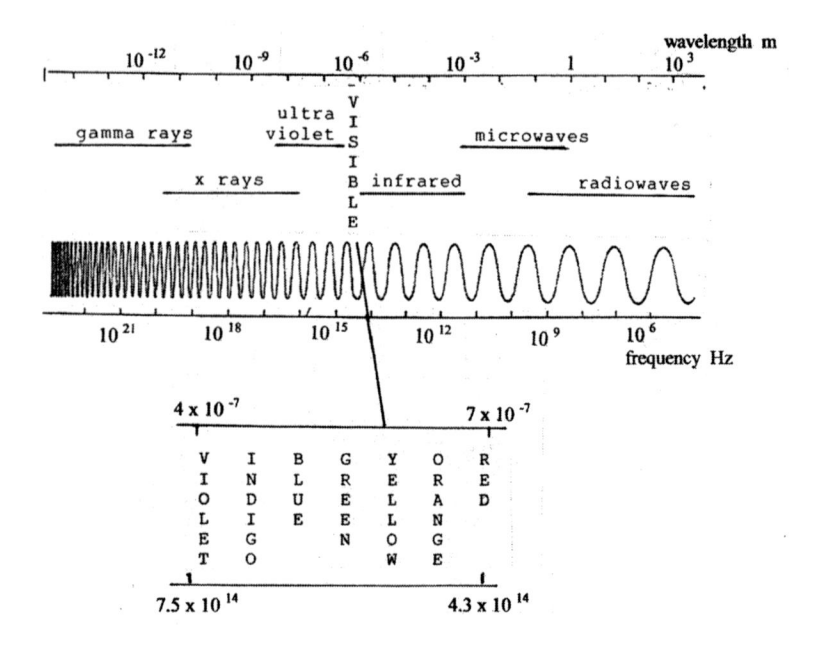

```
         wavelength m
10⁻¹²      10⁻⁹       10⁻⁶        10⁻³       1        10³

                      V
              ultra   I
  gamma rays  violet  S
                      I
        x rays        B  infrared           radiowaves
                      L
                      E

    10²¹      10¹⁸      10¹⁵       10¹²      10⁹       10⁶
                                                     frequency Hz

         4 x 10⁻⁷                      7 x 10⁻⁷

         V   I   B   G   Y   O   R
         I   N   L   R   E   R   E
         O   D   U   E   L   A   D
         L   I   E   E   L   N
         E   G       N   O   G
         T   O           W   E

         7.5 x 10¹⁴                   4.3 x 10¹⁴
```

FIGURE 2 – THE ELECTROMAGNETIC SPECTRUM

Also, just as there are no sharply defined boundaries between each of the types of electromagnetic wave, for example, between infrared and microwaves or microwaves and radiowaves, there is a sort of melding of one into the other, also there is no known limit to the electromagnetic spectrum. It just gets to the point where we can't measure it anymore using present day technology. This factor means that it is entirely possible that the electromagnetic spectrum is a subset of a greater spectrum which I call harmonic multiples of the speed of light. I have already said that the spin of particles is linked to their frequency so this means that spin speed of particles must also increase along with frequency and this is elaborated later in the text. This means that matter can exist in other dimensions because the spin of the particles would be in synchronisation with the faster frequencies but would be outwith our perception in this dimension which is based on the speed of light at 300,000 km per second. The theory presented in this book explains that when spin and frequency are increased in sync beyond the known electromagnetic spectrum, then faster speeds of light are accessible and so matter can exist at these faster speeds. This is the 'missing matter' or 'dark matter' of physics.

Quantum physics

The business of reducing everything to its smallest constituent is known as reductionism and has given us the theory of quantum physics or quantum mechanics as it is often called though I prefer to call it quantum physics as 'mechanics' sounds too devoid of meaning and too much like the Newtonian mechanical model of the universe which has been dominant for the past 300 years, which I doubt that Newton himself would have subscribed to. He is credited (or discredited) with having described the universe as a giant clockwork mechanism but as I refer to later, I believe this is a misrepresentation. If he did describe the universe in this way, he also believed that the clockwork was designed and kept in order by a Creator. This gives a whole new meaning to the mechanical model. However, because of the associations of the word 'mechanics' I shall refer to the theory as 'quantum theory' or 'quantum physics'.

The word quantum comes from the study of how the quantas of energy move (a quanta being a discrete unit or quantity of energy proportional to frequency of radiation, or a share, a portion or an allowed amount, according to the Concise Oxford dictionary). The theory of quantum physics tells us that the sub-atomic particles seem to appear and disappear, they seem to exist in several dimensions at once and that the observer of the experiment affects how the particles behave, simply by observing them. Existing scientific laws simply cannot explain these phenomena. One of the basic tenets of scientific experiment is that the observer must take every precaution not to affect the outcome of the experiment by controlling the variables. The variables are the outside influences which affect the results of the experiment. In quantum physics, of course, we know that the observer <u>does</u> affect the experiment, so the observer then becomes a variable. By observing, a particle appears to change behaviour. This is the same as saying the observer is creating, or changing reality, simply by observing. This has profound implications for the whole scientific model and our view of our place in the universe. But it is also showing that consciousness is light, that consciousness is creative and that light obeys the laws of each dimension. This is elaborated in Chapter 14 – The Law of Resonance. It also suggests that there is no such thing as true objectivity. It shows that subjectivity affects objectivity. Now even though quantum physics is showing us that this is the case, this thinking is still not mainstream science and institutions are still teaching that objectivity is the true model and subjectivity has no place. Yet here is a direct contradiction.

Non-locality

Another contradiction in science is found in non-locality (this means not confined to one place only but is everywhere). Quantum physics has shown that sub-atomic particles display this quality. The implications of this mean that time and space are illusionary or elastic. This means that sub-atomic particles display omnipresence (can be everywhere). Since everything arose from an original source, some call it the Big Bang, our bodies, and everything in the universe are made from the same original ingredients. This means everything is non-local, omnipresent or connected instantly. This means the known speed of light cannot be the only speed of light. If a particle here on Earth is in instant communication with a particle in the farthest reaches of the galaxy, then there must be speeds faster than the known speed of light, so the known speed must be relative.

Another contradiction is Heisenberg's Uncertainty Principle. This means that it is impossible to determine both the position and momentum of a particle simultaneously because when you illuminate (observe) a particle you change its momentum. Mainstream scientists argue that these two laws (non-locality and Heisenberg's Uncertainty Principle) don't apply to human consciousness but it will be explained later that they do, since, it will be argued, as consciousness is light and light is non-local, both laws apply to human consciousness. Other experimental work has demonstrated that quantum events such as these are indeed operating in human consciousness [1]. The proof for non-locality is presented in Chapter 22 - The Heart and Quantum Concepts.

Quarks

The sub-atomic particles of the nucleus of the atom, that is the protons and neutrons, are believed to be made of even smaller particles which have been named quarks, which, due to their differing qualities have been given further exotic names such as 'charm', 'strange', 'beauty' . This is an example of unnecessary complexity mentioned in the introduction and reeks of pretentiousness. The physicists are introducing more complex laws to try to explain these differing behaviours of the protons and neutrons but what they call quarks are most likely due to electromagnetic interactions between the protons, electrons and neutrons. Their differing behaviour is because no particle is static, each and every particle is in constant motion because it is a particle of spinning lightwaves and it is the dynamism of the constant motion which creates differing behaviour. And no-one has actually seen a quark, they are simply a theory, and a very complicated theory at that.

Baryons

In high energy research in particle accelerators, which incidentally, costs the taxpayer billions of pounds or dollars, some new particles were apparently discovered which have a greater mass than protons but which have a very short life. Protons are very stable particles, their lifespan is estimated at 10^{33} years, whereas these new particles being created in the accelerators have a lifespan of approximately one ten billionth of a second. Because these particles have a greater mass than protons they have been named baryons. Baryon is Greek for heavy (heavy due to their mass). But they have been found to decay into protons with a release of energy which decreases their mass. A whole new law was created to account for the different lifespans of protons and the other heavy particles which is called the law of Conservation of Baryon numbers. This law describes mathematically that these particles behave in a certain way but no explanation as to why. The law sounds very complicated and intellectual but really doesn't explain very much about why things are the way they are, whether these particles are natural or artificially created for example. It is an example of trying to fit new discoveries into the existing laws. There are many names being given to sub-atomic particles - quarks, kaons, muons, gluons, pions, hadrons, leptons, gravitons, etc. some of which have never been seen and I suggest you take a tranquiliser before looking up the meanings in the physics dictionary. Pay particular attention to the phrase 'believed to exist'.

What's happening in the particle accelerators is that particles are being forced into existence very briefly but they radiate away back into a wave again because their creation is temporary. The reason they are creating all these new heavy particles, such as baryons, is because force is being used. They are accelerating particles and in physics we know that greater speeds increase mass. Mass is nothing more than stored energy, or light. So the mass they accumulate is the accumulation of lightwaves just as an astronaut travelling at increasing speeds would accumulate greater mass. Similarly with mesons, created from the debris when cosmic radiation strikes matter. Their life span is a few millionths of a second and they are simply the released light which has been accumulated through speed which is released on collision with matter. But as we will see in Chapter 13 - Morphic Resonance of Light, increased accumulation of lightwaves eventually creates a translation into a higher harmonic dimension of the speed of light what some may term a fifth dimension.

So, many of what's being called 'new' particles are simply particles which have been artificially created due to speed. They are creating artificial particles with an accumulation of lightwaves from a faster dimension (because speed is the access key to faster dimensions). If anything, these experiments show that particles, hence atoms, hence matter, is really non-substantial, fluid and displaying ethereal and ephemeral qualities. It is also showing that particles obey the laws of each dimension. When particles are speeded up they accumulate more lightwaves (because the faster dimensions both vibrate and rotate faster) then when the particles return to their normal speed of this dimension, they throw off the excess light to return to their normal state. This is same phenomenon observed in biology, called 'delayed luminescence' where a particle is subjected to intense light which it initially absorbs but then after a time, radiates it away to return to its normal state because Nature is always trying to maintain homeostasis, which is a relatively stable equilibrium, or relative balance in other words.

Creation of Neutrinos

Science states that the positive charge of the proton attracts an electron and through something called the weak nuclear force creates a whole new particle called a neutron. It also states that the proton contains two up quarks and one down quark and the newly created neutron contains two down quarks and one up quark. It also states that the electron then becomes something called a neutrino. The neutron has a slightly higher mass than the sum of both electron and proton and through a process called beta-decay, falls apart again into an electron and proton. But science insists a neutron is not a bound electron and proton. Confused?

Like so much of science the quark theory has been made over-complicated. Without meaning to sound cynical, perhaps to continue the supply of funding and jobs and it is interesting that the word 'quark' is German slang for 'nonsense'. Going back to Nigel Calder's statement in the Introduction, by 'probing into the half-hidden electronics of the universe and short-circuiting some of its basic components', in this case the weak nuclear force, the theory becomes very simple. The weak nuclear force is used to describe a joined electron and proton becoming a neutron but this can be explained with morphic resonance, inherent in which is electrical attraction. So it is likely that it is electrical attraction which is creating the differing qualities of the proton and neutron, and the up and down configuration of the 'quarks' is suggestive of an electrical connection, the 'quarks' themselves probably just being electrical charge. The fact that the particle is spinning means that it will be creating electrical charge because spin creates charge. The strong nuclear force is the force which is holding the protons and neutrons together in the nucleus. Both the strong and weak nuclear forces can be explained by morphic resonance and will be elaborated on in the chapter on Gravity and the Strong and Weak Nuclear Forces. Morphic resonance means they are held together by a pattern or a template which is resonating. **The resonance or sound is what creates the forces and the particular structure of the atom.**

The interaction of particles in the atom is due to electromagnetism. The positively charged proton is attracted to the negatively charged electron. They join creating a third force. As with all male and female reproduction, the joining together creates a third life, or more life. They join together creating what's called in the business, a neutron, but which is really just a bound electron and proton. The neutron has a slightly higher sum mass than the two individual particles, accounted for by the third life. The neutron, after creating this new life, decays back into an electron and proton and radiates away the third life (which accounts for the extra mass of the neutron). Science calls this the neutrino which they say is the transformed electron. I would argue that it isn't the electron but is the newly created life – the neutrino baby. Then through a process called beta-decay the neutron decays into an electron and proton, but science says the neutron isn't a bound electron and proton! If the electron and proton join together to create the neutron and the neutron decays back into an electron and proton, logic would suggest that the neutron is a bound electron and proton!

Neutrinos exist everywhere, passing through us all the time. Science is presently trying to measure them to see if they have mass as this may account for the so-called missing mass in the universe. Neutrinos are continuously being created because life is continuously being created and this is how it is done, on a particle physics level. This can explain the expansion of the universe if indeed it is expanding. Neutrinos are consciousness (or light) waves which eventually become particles due to the part they have to play in the resonance of OM. They also travel in a spiralling motion when they are created, consistent with other lightwaves. They are the baby of the protons and electrons. Due to the electrical attraction of protons and electrons there is continuous creation of life (neutrinos) and continuous decay (annihilation of electrons). So the whole system is dynamic, continuously creating by the attraction of polarities, but with decay happening at the same time, such as in the growth of any organism.

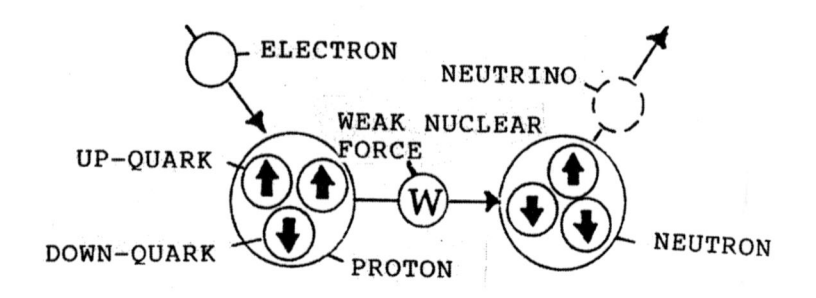

```
         ○ ─ ELECTRON
                        NEUTRINO
              WEAK NUCLEAR
              FORCE
UP-QUARK   (↑↑)──(W)──▶(↑↑)
           (  ↓ )       (↓↓)     NEUTRON
DOWN-QUARK          PROTON
```

*FIGURE 3 – ELECTRON/PROTON BECOMING
A NEUTRON/NEUTRINO*

The name neutron suggests that it is neutral, meaning without electric charge and this is largely the view presented by science. However, this is not strictly true. Although the positive and negative charges of the proton and electron mostly cancel each other out, they do not totally. The neutron has been found to have an electric dipole [2], meaning a negative and positive charge on opposite sides creating a little magnetic field around it, just like a planet. The fact that it has been found to have a magnetic field, of –1.91 nuclear magnetrons means it must have charge because magnetism is created from the spin of charge. This means it is a little electromagnetic particle. As explained in the chapter on The Golden Ratio and Fractals, this equation of electricity to magnetism may be necessary to create life. This would enable the little neutron to create the neutrino. It hasn't been proven, but it could be that the ratio of the electricity to magnetism of the neutron is the golden ratio, sometimes also called the golden mean, which is the ratio of 1-1.618, and which is significant to the growth of form. This would make interesting research.

Conservation of Energy Law

Another seeming contradiction in science is that of the expansion of the universe. If this is found to be correct, then it would imply that more energy is being created which means the Conservation of Energy Law cannot be correct, universally. It only appears to be conserved within the physical dimension but there are many anomalies which suggest otherwise, such as the survival of consciousness when the body dies, when healing takes place and the phenomenon of zero point technology. The faster-than light-speeds theory presented here means that the Conservation of Energy Law must be changed/expanded. It only works within the physical dimension when the above anomalies are ignored. We don't know if energy is conserved in the 100% of reality and we don't know how many dimensions there are. It seems that energy can neither be created nor destroyed within the physical dimension, but within the 100% of reality, of which we are only aware of a tiny fraction, it seems that it is being created continuously. It occasionally leaks through the veils of the faster dimensions and can be tapped and this is when many orthodox scientists close their eyes and ears. This is what is happening with healing and with free energy or zero point technology. It is called zero point energy because even at temperatures of absolute zero, which is the lowest possible energy state, it is still possible to detect fluctuations in the field of space. These fluctuations, which have been known about for a long time, were found to be ever-present in the emptiest state of space at the lowest possible energy state. Because they were ever-present, physicists decided it didn't change anything and so they (the fluctuations) have always been calculated out of physics equations. This is called in physics 'renormalisation'. This process of renormalisation has allowed science to dismiss what is 'the essence', the 'missing ingredient' in not only reconciling Relativity theory with quantum physics, but also in explaining many seemingly unexplainable phenomena, such as consciousness, non-locality, telepathy, clairvoyance, healing, remote viewing, zero point technology, to name a few. In zero point technology, the energy appears to be coming from nowhere but it is simply coming from the fluctuations, which is energy, or in this theory, light, in the zero point field, or another way of saying it is, another dimension at a faster speed of light It may be that the zero point field which is measurable, may only be the next dimension up. In other words, the fifth dimension, or fourth if you don't include time. In the theory presented here there would be many higher dimensions, the number is unknown, but there is no way of measuring them directly at present, although their *effects* can be measured, such as the experiments with remote viewing, telepathy, zero point technology, prayer and healing. The latter two experiments are described in Chapter 22 - The Heart and Quantum Concepts.

The zero point field

The zero point field is known about by quantum physicists and it is also accepted that there is no such thing as a vacuum or a state of no-energy. Yet even today, there are still books and papers being written by scientists, including quantum physicists, talking about the vacuum of space! It seems that it takes a very long time for new findings to filter into consciousness or for language habits to change, even amongst scientists and this keeps us stuck in outdated understandings. **Another aspect of Heisenberg's Uncertainty Principle is that no particle can ever be truly at rest because the fundamental structure of the universe is in constant motion; a vast ocean of fluctuating quantum events. But this is further proof that it is all light because light has no rest energy, it is perpetual motion.** It seems incredible that for so long no-one (at least in mainstream science) thought these fluctuating energies might be relevant. The higher dimensional energies have always been an established reality to many people who tap into it regularly, in meditation or for healing etc. They just have a different name for it. But one scientist who did think the fluctuating energies might be relevant was Timothy Boyer of City University of New York, who published a paper in 1969 in which he described how classical physics could be reconciled with quantum theory if the zero point field were included in the equations ³. But perhaps the world wasn't ready to hear that yet. Probably the main reason why mainstream science is still digging in its heels is because it doesn't fit the known Conservation of Energy Law. This law, as presently understood, applies only to the physical dimension which is based on the known speed of light. **The fact that the zero point field is ever-present and doesn't change the physical laws tells us two things a) that because physical equations are not affected by it means that it is beyond physical, in other words, a different dimension, and b) the fact that it is ever-present and beyond physical means that there must be other laws to describe it. So the laws of physics need to be expanded to include the other dimensions.**

The world of the individual atom is thought to resemble a tiny solar system, an example of 'as above, so below'; the nucleus being the sun, surrounded by circling electrons (like planets). Like the solar system most of the atom consists of space, which means all of us are made of mostly space. This space, however, is not empty. It is full of energy, though not apparent to the five senses. The space is spiralling light waves of consciousness, what I have already described as the ether, zero point field or dark matter or the 'strings' in string theory.

String theory

String theory is another increasingly popular theory in science which recognises that there are multiple dimensions, at least 11, in which tiny 'strings' vibrate. However, string theory can't describe what the strings are made of or how they form sub-atomic particles, or how form takes the shape it does or how the other dimensions are accessed. It is a very incomplete theory as yet but the 'strings' part of it is consistent with the theory presented here in that the 'strings' are the spiralling lightwaves existing in many dimensions, which vibrate to a particular resonance and so become matter. If you remember that sound creates matter from lightwaves. Strings are an interesting choice of name for the theory and very apt. Just as strings in an orchestra vibrate to different frequencies, producing the overall harmony, so do the 'strings' of the smallest particles of matter resonate to frequencies of the overall harmony. **Morphic resonance of light is an orchestral harmony.** Further evidence of these vibrating, resonating strings is from the Vedic rishis. The Vedas, four in all, were written in Sanskrit, the ancient language of India, between approximately 1500 and 500 B.C. and remain the spiritual foundation of Hinduism. They say the cosmos is made of sutras, which is their word for string, stitch, thread or verbal phrase. This is no doubt where our word suture comes from. So here we have a perfect dual meaning of sutra, a string or a verbal phrase. This is a perfect description of the spiralling lightwaves (strings) resonating to the sound of OM (verbal phrase). The Vedas liken the cosmos to a tapestry – a pattern made out of verbal phrases. This is a perfect description of the morphic resonance of light.

CHAPTER 4

THE DISHARMONY

> *"We are no longer inheriting the Earth from our parents, we are stealing it from our children."*
> David Brower, Environmentalist, Conservationist, 3 times Nobel Peace Prize nominee (1912 - 2000)

This study of these miniscule particles of the sub-atomic world, which could be described as reductionism, (reducing the study to smaller and smaller pieces) is often derided today in the increasingly popular quest for holism, or holistic ideas and concepts, the interconnectedness of everything. Holism is a necessary aim and one which will ultimately save us from our self-destruction. Mankind today is poised on the brink of self-imposed destruction due to the misuse of free will. As noted earlier, we have the free will to use science and technology at our behest, but if the intention is driven by profit or even the ignorance of interconnectedness we will see the results. And we do see the results today. We now have a situation in the world of gross instability in all aspects of life; a disequilibrium, a disharmony. This has become very obvious within what we call 'the environment'. We have become disharmonious, out of harmony, with OM. In other words, we have become out-of-tune within the great Harmony of the Spheres, the OM. We have to understand that all life is interconnected so for example, what we do to the environment affects us, either directly or indirectly. For example, what food we choose to eat has an effect far beyond the vitamins and minerals we absorb. It is necessary to think about where the food came from: the pollution caused by the distance travelled from source of growing to your plate, such as aeroplane fuel, shipping fuel, truck or van fuel on delivery to the supermarket, car fuel from the supermarket to your home, the electricity or gas used to cook it. The chemicals used in its growth; such as pesticides, herbicides, fungicides, artificial fertilisers and the impact they have on the soil, on the people using them and on you after you eat them. Also, the packaging of the food: plastic wrapping which is produced from fossil fuel then has to be disposed of, creating landfill waste or even if recycled still uses fossil fuel to do so. Also plastic wrapping is known to contain harmful chemicals which can leach into the food. Some of these chemicals display xenoestrogenic behaviour, which means they behave in the body like oestrogens and as we know, synthetic oestrogens or even an excess of natural oestrogens can promote cell growth and are implicated in cancer development.

Animal welfare is another factor to consider. As will be explained in more detail in Chapter 24, the way animals are farmed and the conditions they are kept in, when produced intensively, is not only cruel to the animal but it determines the quality of the meat, which affects the person consuming the meat. This is to do with the light energy, or biophoton energy which living things radiate (which can actually be measured) and is affected by living conditions, diet, medications etc. There is a health and environmental price to pay for intensively-farmed, cheap food, either animal or plant. These ideas have been resisted for a long time but are becoming more and more accepted as the information is becoming more widely available. The information, however, has been around for a long time. People like Lady Eve Balfour, Friend Sykes and a few others who founded the Soil

Association around the end of the Second World War and Rachel Carson, author of 'Silent Spring' published in the early 1960's, spoke passionately on this subject.

As long ago as the beginning of the 1900's a young scientist, Albert Howard, mycologist and agricultural lecturer to the Department of Agriculture in Barbados, was discovering that in order to know the true cause of plant disease and how to remedy it, it was necessary to be out in the field actually growing and tending plants rather than study specimens through a microscope in the lab which was where all his work was being conducted. As he discovered "there was a wide chasm between science in the laboratory and practice in the field". In 1905 however, his chance came to combine both when he was appointed the position of Imperial Botanist to the Indian Government, based at the Agricultural Research Station in Bengal. By closely observing the native agricultural methods he discovered they were producing healthy plants and animals without any artificial chemical pesticides or fertilisers. They were achieving this by simply returning to the soil composted animal and plant natural waste. The health of the soil was everything, and this is the basis of organic farming today. Healthy soil, or a healthy environment creates healthy immune systems. This was not just a theory for Howard as he managed to prove it from his own experience and the startling results he achieved have great relevance to us today, in that his herd of oxen, reared on his fertile land, resisted every (frequent) outbreak of foot-and-mouth, rinderpest and all the other cattle diseases which surrounded his herd. Not once did he segregate his herd. They were living cheek-by-jowl (literally) as he often saw them rubbing noses with diseased animals, yet they never succumbed to disease. He concluded that "the basis for eliminating disease in plants and animals was the fertility of the soil". This, he discovered was also an ancient Chinese practice and no doubt of many other ancient cultures and he wrote up his findings in a book entitled "The Waste Products of Agriculture: Their Utilisation as Humus" which received praise from around the world. Howard was perhaps the main inspiration behind Lady Eve Balfour's interest in the subject. She used his methods and cured herself and her animals of disease, and went on to write a book which documented her findings called 'The Living Soil' published during the Second World War.

Albert Howard went on to be knighted for his accomplishments but alas there were many resistant to his discoveries, mostly research scientists and chemical companies who made the pesticides and fertilisers, as you might expect. Same old problem as noted in Chapter 2. For the scientists, his findings were seen as a threat to prevailing world view, self image and perhaps professional jealousy. For the chemical companies, it meant loss of profits. The whole sorry story is reported in 'The Secret Life of Plants' by Peter Tompkins and Christopher Bird, published in 1973 by Harper & Row, U.S.A., which remained on the bestseller list for several months at that time. But even Charles Darwin (1809-1882) was a fan of healthy soil and recognised its importance to a healthy ecology. He published a book in 1881 called 'Vegetable Mould and Earthworms' in which he described the importance of the presence of earthworms in particular to a healthy soil, claiming that earthworms are crucial for plant life. Vegetal and animal waste returned to the soil, or compost as we tend to call it these days, is the main food for earthworms and in soil where only artificial fertilizers and pesticides are used, earthworms can disappear entirely. So this knowledge has been around for a long time but just not promoted enough, or perhaps been suppressed. However, now thankfully, these ideas are beginning to creep into the mainstream.

Everything we do has an impact on so many other things so to live holistically means to be aware of and cautious of the impact your actions have on everything else. With the rising awareness of pollution problems and the depletion of fossil fuels it has become socially acceptable at last to think, speak and live holistically. It will become the norm for all before too long. It will have to become so if we are to survive as a species.

Living holistically doesn't mean feeling guilty about the impact you are having on the planet, nor does it necessarily mean having to give up comfort, it simply means to be aware of the interconnectedness of all life which brings with it a caution in one's actions but also a sense of the sacred, a reverence for all life and a deep and profound appreciation. It is being able to see the interconnectedness of all things and living in a respectful way of that.

Many people seem to rail against these types of ideas because it provokes guilt in them or a feeling of not wanting their comfort zone to change or a feeling of resentment towards the person living holistically because the assumption is that they somehow feel superior to those not living holistically, but this feeling is usually based on prejudice and misperception, which usually comes from wrong education. There is a whole section of society who believes that living holistically is only for the middle classes and upwards, especially since it involves eating organically grown food, which sometimes costs more. This is another example of the 'them' and 'us' scenario mentioned in the Introduction and explained in the next chapter, which is due to duality consciousness and of which humanity is ensnared within. While it's true that some supermarkets charge extra for organic food, it is not always more expensive and there are many box schemes around the country where the produce is delivered fresh each week from a local farmer at very competitive prices. Farmers' markets are also becoming very popular. These are beneficial trends as they cut down on transportation pollution. It's common sense that there will be less CO^2 emissions released into the atmosphere by locally sourcing foods rather than transporting them halfway across the planet. As well as this benefit, the food is fresher so has not degraded so much as it would if it had come from a further distance. Also, as will be explained in Chapter 24, there is a health price to pay for eating food which has been intensively grown, because of the chemicals involved. As already described, a healthy soil produces a healthy immunity. This is both literally and metaphorically true.

There are health and economic benefits to living holistically, both for the individual and nationally and much that the individual can do to begin the process, for example, by growing some of your own food and composting vegetal waste, or using the farmers' box schemes, or refusing to buy over-packaged goods, or by lobbying your supermarkets or MPs for more eco-friendly packaging, and a thousand other ways. There are several good books around with helpful advice on this subject.

So there are ways and means to live holistically if the will is there. There is a spiritual maxim that says 'if you want to change the world, begin with yourself', or as Ghandi said "be the change you want to see in the world". The relevance of this will become clear as we go on to talk about the fractal or holographic nature of the universe and the principle of resonance.

CHAPTER 5

HOLISTIC CONSCIOUSNESS

"No problem can be solved from the same consciousness that created it."
Albert Einstein, Physicist and Mathematician (1879 – 1955)

Reductionism, however, has served a purpose. As it applies to quantum physics it has led to great surprises, and as is often the case with great mystical truths, looks like a paradox. The paradox is that the reductionism of quantum physics has led to a discovery and strong evidence of holism, which to spiritually aware people, has always been a self-evident truth. This is because in particle physics, the sub-atomic particles seem to exist everywhere 'potentially' at once and photons (particles of light) are continuously exchanged between electrons. So it really shows a holistic interconnectedness and dynamism.

The paradox, as it seems, is only due to the level of consciousness in which most of humanity operates. That of dualism, or dualistic thinking. Spirit vs. matter, reductionism vs. holism, science vs. religion, them and us. Dualism is a limitation to understanding. What is required is dialectic thinking. This means a process of integrating opposite extremes of logical thought in order to come to an understanding of the underlying unity behind the apparent opposites. The German philosopher George W.F. Hegel (1700-1831) described dialectic thinking as "the process of thought by which such contradictions are seen to merge themselves in a higher truth that comprehends them."

As humanity evolves, the diametrically opposed views come towards a central point, in other words, a dialectic view. Humanity at the moment, however, is so split in its views that we not only have the ridiculous scenario of science versus religion, but within both sides we have disciplines or ideologies that won't speak to each other. How can we possibly have a unified theory of the universe, which is the great goal of science, if we cannot even communicate with each other?

This is not to say that opposites are somehow wrong; they are a fact of life, but the point being made is that humanity does not always honour the opposite as an equally valid reality or see that both may be part of something bigger and in addition, has further divided itself up into different camps within the opposites, creating more and more fragmentation. This has happened with both science and religion. We all know that the best diplomats are people who can understand opposing points of view but takes neither side, or perhaps takes both sides. From the central point both sides can be understood equally. Opposites, as already explained, are part of something bigger. The whole of life on Earth is based on dual principles. The male and female genders for example are found in every kingdom, from the mineral, through plant, animal and human. Even within each person we have so-called masculine and feminine qualities. Our brains are composed of two hemispheres. The left hemisphere controls the right side of the body and is referred to as the masculine side because it deals primarily with what are commonly described as masculine attributes; expressive, linear, analytical energies. The right hemisphere controls the left side of the body and is referred to as the feminine side. It deals primarily with receptive, spatial, intuitive, energies. This is a generalisation and the masculine and feminine should be viewed not as gender issues but as complementary forces, as with electricity for example. Now the fact that we all have two sides to our brain means we should be holistic individuals who have all of the above qualities operating synergistically. We would then be balanced beings. However, as we know this is not the case with most people.

For at least the past two thousand years male energy has been dominant, creating a very patriarchal and unharmonic society which has fuelled the dualistic gender divide. This has been due in part to the energetic influence on our planet of the Pisces constellation but this is changing now as we are moving into alignment with the constellation of Aquarius. I am aware that to mention astrology causes eye-rolling in many people, particularly the scientific community, who find it very hard to take the subject seriously. Not surprisingly perhaps when we see the popularisation of horoscopes in the daily tabloids, much of which is superficial nonsense. However, astrology *is* an ancient science and many scientists of the past were aware of the influence of astrological alignments. This is quite natural when a person has a holistic understanding of the universe and an awareness of an energetic interconnectedness between all matter and space. This is now the findings of quantum physics and those at the forefront of this science are now seeing that matter and space are not what we previously thought they were. One of those scientists of the past was the astronomer Johannes Kepler, who in 1618 published his meticulous study into the movement of the planets, called '*Harmonices Mundi*' or '*The Harmonies of the World*'. Kepler was also a court astrologer and dedicated this work to James I of England. However, with the outbreak of the Thirty Years' War his work fell into obscurity and somehow over the centuries astrological knowledge fell out of favour with mainstream science. The reductionist, separative mindset has been so dominant that subjects like astrology have become the subject of ridicule. But if you think of the influence the Moon has on the Earth, creating the tidal patterns, and the fact that the Sun emits positive and negative ions in seven day cycles which affects electromagnetism on Earth, not to mention sunspots which also affect Earth's electromagnetism, then it's not too difficult to see how astrology is the study of energetic influences between bodies in space, remembering that human bodies are also electromagnetic.

The work of Psychologist, Stan Grof and Physicist, Rick Tarnas, at the Esalen Institute in California is worth looking into in this regard. They have discovered correspondences between astrological alignments and transformative events in Earth's history. This information is not new and has been documented throughout the ages but has not been given prominence. Hopefully, this will be corrected with the seminal work of Stan Grof and Rick Tarnas. In astronomy we know about the precession of the equinoxes, meaning that as our solar system travels around the heavens, it comes into alignment with the different constellations of stars. There are 12 constellations, each lasting approximately 2,100 years duration, resulting in a 26,000 year cycle. This is the length of time it takes for the 360 degree cycle of the Earth's axis tilt; what's known in astronomy as the Chandler wobble. So Pisces has come to an end and with it the end of its unique energetic influence. It should be said that this is a gradual change over many years, there is no fixed date to the changeover. The difference is in the degree of radiation, analogous to the electromagnetic spectrum, where there is no solid definite line between one type of energy than another (e.g. between microwaves and radiowaves). Aquarius is reported as being a more harmoniously balanced energy between masculine and feminine so this is good news. However, it should be said that astrological alignments in no way absolve us from our own personal efforts to be balanced individuals.

In the human brain, the corpus callosum is a collection of nerve fibres which adjoins both hemispheres of the brain and sits at the top of the cerebellum. It connects both hemispheres through nerve impulses and, in essence, lets each side 'talk' to the other, creating a holistic view, or synergistic understanding; seeing the bigger picture and not just one viewpoint. It is sometimes obvious when a person is operating more from one side of the brain than the other. Also, there are certain conditions, such as dyslexia, which may be helped by the simple exercise of tracing a figure eight with both eyes, always crossing over at the same point in the middle. This has the effect of stimulating the nerve impulses of the corpus callosum, strengthening it in other words, which enables better function in the more sluggish side of the brain and enables a synergistic expression of the whole brain. The corpus callosum has been found to be thicker in women [1] and perhaps this explains why women often, though not exclusively, have a more holistic view of life, despite the predominance towards linear, analytical, technological expression throughout the educational system and society in general. This predominance has resulted in denial of the spatial, intuitive side which would create a more balanced and holistic individual. Some men, however, are also quite obviously expressing balanced and holistic qualities. Perhaps it means that their corpus callosum is thicker than other men's or perhaps has better nerve function, and vice-versa for women who seem deficient in holistic understanding. Is their corpus callosum thinner than the average woman's or does it have depressed function? This would make interesting research. The point is that both sides are necessary to create a whole understanding and expression; a well-balanced individual.

But even just considering the physicality of the body, it is now known, through the study of what's become known as psychoneuroimmunology, that the brain communicates directly, biochemically, with the cells of the body. This has obvious implications for the body if the brain is only half-functioning or one side is being suppressed. This correlation is often seen in the healing arts, when, for example, a masculine energetic imbalance creates a symptom in the right side of the body, which is the masculine side, because the left side of the brain controls the right side of the body. (Remember we are talking here about energetic influences and not gender issues). There is a definite mind-body connection which has always been an established reality in spiritual philosophy and to those who practice the healing arts. In modern, conventional medicine the mind and body were, until quite recently, viewed as distinctly different things with no connection at all, but with the discovery of psychoneuroimmunology it was discovered they are part of a whole, that the mind, particularly the emotions, creates chemical interactions with the glands and the immune system. This is why visualisation for treating various ailments of the body works so well and why healing methods like emotional freedom techniques, originating from acupuncture, and brought to prominence by Dr. Roger Callahan and Gary Craig, can have such dramatic healing outcomes. It is thought by some in the medical profession that perhaps more than 80% of diseases are caused by suppressed emotions. The figure could be disputed, it could be lower or higher, but many are now realising the impact of unexpressed or suppressed emotion on health. The physics of this are explained in Chapter 24 The Cause of Cancer.

One of those at the forefront of the understanding of psychoneuroimmunology is neuroscientist Candace Pert, Research Professor at the Department of Physiology and Biophysics at Georgetown University Medical Centre, Washington, D.C., whose pioneering research back in 1972 led to her discovery of the opiate receptor, providing irrefutable evidence of the communication between emotions and the immune system. In her book, *Molecules of Emotion*, she outlines her adventure to bring this revolutionary discovery to scientific awareness and acceptance. She ran up against the old problem of orthodoxy when presenting her discoveries and encountered, but overcame, deeply entrenched sexism from male colleagues, which, unfortunately, still exists within the Establishment of orthodox science. She cited the case of Rosalind Franklin, who had discovered the critical chain of reasoning which enabled Francis Crick and John Watson to show that DNA was a double helix, for which they received a Nobel Prize.

But what is not so well known is that they persuaded her boss to let them see her findings while she was away from her lab (some would call this theft !). Although she was briefly acknowledged in their published paper she was not included in the Nobel Prize. Worst still, Watson actually boasted about the theft in his book and proceeded to speak of her in a derisory manner. Rosalind Franklin died a few years later from cancer, never having had her voice heard. In light of the understanding of the effect of unexpressed emotion on the body, outlined in Candace Pert's book and discussed in the chapter on The Cause of Cancer, it is perhaps not surprising that this took her life. Viktor Schauberger's death from a broken heart is another example of emotions creating bodily symptoms. He was already in a weakened state before making the trip to the U.S. and the overwhelming emotions he felt about the direction science was taking most likely 'broke his heart'.

Candace Pert's work is now becoming well-accepted and is a perfect example of holistic, mind and body interconnectedness. Other pioneers of mind/body interconnectedness and how this can be used for self-healing, such as Louise Hay, have taught these concepts for a very long time. Though, as with so many other things, there is still some resistance to this discovery. Due largely because it was believed that the mind, especially emotions, were considered entirely separate from the body with no influence at all on the immune system and there are still some in the medical profession today who conduct their practice within this assumption. These medical people are similar to the scientists who still talk about the 'vacuum' of space. So we can see that it is because of the reductionist and separative approach of science and particularly the predominance that has been given to the left side of the brain, that we have such a fragmented view of reality.

As well as a disharmony between the two sides of the brain, there is a further disharmony between the brain and the heart. Humans are quite compartmentalised in their functioning and we are not yet functioning holistically. We have energy centres up and down the body (the chakras, described in Chapter 23 Auras and Radiation) which, until they are flowing freely, will not produce a true expression of our potential as not only intelligent beings but also unconditionally compassionate beings. Compassion and human values tend to be expressed from the heart centre . In many people the heart centre and the right side (feminine) of the brain are suppressed. The left side of the brain rules. Society, particularly in the Western world, worships at the feet of the God called Commercialism, whose minions include market forces, the economy, technological and military supremacy. These are the new Gods of planet Earth. They are evidence of a completely lop-sided approach to life. They need to be balanced with a holistic understanding and an opening of the heart. Our technological development has not been matched with our spiritual development, and this has created a very dangerous situation in the world. The whole brain and heart need to be deployed to bring us up to speed in our spiritual development so that we can responsibly run this planet for everyone's benefit and avoid annihilation of all life. The universal values of Truth, Love and Peace, taught by 'sages through the ages' need to be taught and lived more urgently now than at any time in the past. We are often said to be the most intelligent species on the planet, but this is questionable. We have created a state in the world of such imbalance, ecologically, socially and politically, that we are now poised on the brink. If we make it as a species now it will be due to the practicing of human values en-masse. This would be the intelligent solution, synergistically expressing from the whole brain and heart. I would argue that compassion and human values are the mark of an intelligent and evolved person, who is operating holistically, so the fact that these qualities should be seen as 'not economically viable' or 'for softies and impractical for daily living' is an indication of the lop-sided reasoning in which society currently operates which, in itself, is an indication of the lop-sided functioning of human brains. To further support this argument of the importance of holistic, synergistic expression, it has been shown in studies of people who practice Transcendental Meditation that in the course of practice, the frequency patterns of the brain, which tend to be different for each side of the brain, come into a coherence, in other words, the brain is functioning holistically or synergistically. Practitioners report heightened awareness, creativity, intelligence and calmness of purpose.

It may not always be necessary to have both sides of the brain at the same frequency, not enough is known about that. As will be explained in Chapter 16 Electro-Magnetism and Chapter 20 The Golden Ratio and Fractals, perhaps evolution proceeds due to the stimulus of one side bringing the other along, particularly if expressed in the golden ratio, i.e. 1 – 1.618. Perhaps there is meant to be a golden ratio of expression from the brain, of one side to the other, but this has yet to be determined, but the problem would be if there was suppression of one side because then there could never be harmony. To be in harmony does not necessarily imply equality, or both sides the same. It just means "an agreeable effect of the appropriate arrangement of parts" which is also a description of coherence. And coherence, if you remember, was described in the introduction as the realisation of the whole, where dualistic principles are merged and can be understood as reciprocal pairs of one whole.

CHAPTER 6

A LITTLE HISTORY LESSON

"We must begin to teach our children and our citizens how to search for a more true history of the world, and encourage them to look for the truth of the present."
Thom Hartmann, Author and Broadcaster

Quantum physics began to be discovered around the beginning of the 20th century due largely to the work of Werner Heisenberg (1901-1976), Neils Bohr (1885-1962), Max Planck (1858-1947), Paul Dirac (1902-1984), Louis de Broglie (1892-1987), Erwin Schrodinger (1887-1961), Albert Einstein (1879-1955), and Wolfgang Pauli (1900-1958). So, for almost a hundred years the concept has been known about, but little understood. As with many great discoveries it has taken time to integrate the truth into everyday consciousness. A well known and important example of this was Nicolaus Copernicus's (1473-1543) heliocentric theory, i.e. the Sun being the centre of our solar system and not the Earth, as was commonly believed at the time. Later, Galileo Galilei (1564-1642) confirmed this theory, but both were persecuted for their findings and it was many years after their discoveries before the world finally accepted this truth. As is the nature of humanity, there are many who, individually or collectively as an institution, stand as obstacles to the truth, for their own reasons. This was the case with Copernicus and Galileo. The obstacle to the truth at that time was the institution of the church who believed that the Earth was the centre of the universe and therefore, the Sun must orbit the Earth. Conditioned beliefs such as these are the biggest obstacle to truth. Conditioned beliefs come from tradition which, if you recall from Chapter 2, is knowledge of the past, or the inability to see things as they actually are (from new discoveries). So, the findings of quantum physics have also been prone to limited comprehension and acceptance due to conditioned beliefs. Even though some physicists are trying hard to understand what quantum physics means and using words like parallel universes, time travel etc., these ideas are still not mainstream and evoke a kind of embarrassed silence when discussed in polite company, which is always an indication that conditioned beliefs have been offended. Conditioned beliefs are many and varied and the findings of quantum physics seem to fly in the face of everything we have been led to believe up to now. We have been led to believe that the Creator (if one believes in a Creator) is separate from us and does all the creating, or that this physical dimension is all there is and if something can't be measured by instruments then it doesn't exist and is not worthy of attention or that the world is made of physical matter and is entirely objective and that subjectivity doesn't count in science. These have been the normal ways of thinking for several hundred years and have been taught down through the generations by science and religion, *despite* the fact that our everyday life experience tells us differently.

However, quietly alongside this there have always been free thinking scientists. These are people who are free thinkers and who are open to inspiration. They are free thinkers in the sense that their thinking is not confined to orthodoxy. Albert Einstein, in his later years, said that in order to be a creative scientist it is better to be a free agent as far as employment is concerned; to be free of institutional politics and orthodox thinking which normally accompanies a traditional university position. He suggested a menial job allows time for study and research, which he himself enjoyed while employed as a patent clerk in the Swiss patent office. He was also influenced by his early education in Germany which was harsh and authoritarian, where no free thinking was allowed and he railed against this, which led to him being labelled by his teachers as a 'lazy and poor student'. He felt stifled by his physics teachers who seemed unable to consider new ideas and ended up teaching himself through independent study.

I have personally found independent study to be the most satisfying method of inquiry. It allows time for pondering and observation and gives freedom from the burden of mathematical equations. In science the mathematical equation should not be the starting point because mathematics, as Einstein noted, is simply a way of writing down the laws discovered through physics. This is what I alluded to in the introduction; the understanding has to come first.

A true scientist has to feel free to peep outside of the box of known laws and be aware that scientific laws, like everything else in life, are in a constant state of evolution. Now throughout history the scientists brave enough to do that have usually been dismissed, ridiculed, even oppressed or sometimes just not given due credit because the world is still in the dark. This has happened throughout history to people such as Galileo Galilei, Nicolus Copernicus, Nikola Tesla, Viktor Schauberger, Johannes Kepler and even to modern scientists such as Jacques Benveniste and Rupert Sheldrake. They are usually called heretics, but heretic simply means 'those who make the choice to be an independent thinker'. And historical evidence has shown that heretics are often correct. Isaac Newton himself, for most of his adult life, lived in fear of being accused as a heretic due to his secret experiments in alchemy [1]. He was a closet mystic simply because it was not safe at that time to be open on this subject. He was simply trying to understand the transmutation of elements. This aspect of his life has been very much suppressed, however, as it didn't fit well with the image that science wanted to portray to the world of the man who gave us the first laws of gravity. He is usually portrayed as the man who gave the world 'Newtonian Mechanics' which implies that he viewed the world as a giant machine, soulless and devoid of meaning. But I believe this is a misrepresentation. He was a man of great depth and insight, who believed in a supreme Creator and that this Creator was the 'spiritual force' underlying all matter and energy. This idea was also held by Henry More, one of a group of Cambridge Platonists, whose writings influenced Newton. The Platonists took their name from Plato who, like they, believed that spirit permeated all things. Albert Einstein also believed in a Creator as a spiritual force and the fact that we are all part of one ordered Whole when he said "A human being is part of the whole, called by us Universe, a part limited in time and space. He experiences himself, his thoughts and feelings as something separated from the rest, a kind of optical delusion of his consciousness. The delusion is a kind of prison for us, restricting us to our personal desires and to affections for a few persons nearest to us. Our task must be to free ourselves from this prison by widening our circle of compassion to embrace all living creatures and the whole of nature in its beauty. Nobody is able to achieve this completely, but the striving for such achievement is in itself a part of the liberation and a foundation for inner security."

Perhaps if Newton had had the words available to him he would have called the spiritual force the zero point energy and the order, the morphic resonance fields. Being a heretic or making a choice which is different to the tide of common belief, can be frightening and/or dangerous. This is another reason why perhaps there aren't more heretics. The fear in Newton's days used to be of persecution and death, nowadays it tends to be fear of loss of career, status, funding or of public ridicule.

CHAPTER 7

IT'S ALL RELATIVE

"There comes a time when the mind touches a higher plane of knowledge but can never prove how it got there."
Albert Einstein, Physicist and Mathematician (1879 – 1955)

Conditioned beliefs, in whatever form, can be dispelled by inspiration. To be inspired means to take in a higher truth, a higher understanding, to resonate to a faster speed of the zero point field. The word 'inspired' comes from the Latin 'inspirare' – which means to take in spirit or take in breath. Spirit and breath are synonymous in Latin, and in other languages, such as Indian, 'atma' (originally from Sanskrit) means spirit/soul and breath and in Greek 'pneuma' means both spirit and breath. In the Bible there are references to God *breathing* life into Adam. This brings us back to OM. Source breathed out OM. It could be said OM is spirit resonating, a part of which is the fluctuating zero point field. As we know the zero point field is ever-present, this means that spirit is ever-present, which means it is within us all the time and ever available to us. Or another way of saying it is, by attuning to faster speeds, our consciousness is expanded. We become closer to understanding our Source.

It has been described previously how the OM is the breath of Source. It is the spiralling light waves of consciousness which take form due to the sound of the breath. So when a person is inspired, they are absorbing light (higher dimensional light) or they are resonating to an extension of the electromagnetic spectrum. If they are in resonance to it, their spin and frequency of particles will be in synchronisation. When they are inspired they can therefore access higher knowledge, greater truths, profound insights etc. as all knowledge is contained in the higher, or faster, morphic resonance fields (this is explained in greater detail in Chapter 13 Morphic Resonance of Light) and artists, musicians, authors etc. *know* that feeling of being inspired. They feel 'taken out of themselves' is how it is commonly reported and time seems to be non-existent. That is because faster speeds alter time. Their consciousness is literally operating at a faster speed. As faster speeds slow time, ultimately faster speeds would stop time, making a state of no-time, so all past, present and future would coexist simultaneously. This is relativity. This cannot be understood from dualistic consciousness levels. Time, therefore, is relative. Relative to the state of consciousness. Einstein himself proved that time is elastic, i.e. not fixed. That is why his theory is called 'relativity' theory. Actually he gave us two theories, general theory of relativity and special theory of relativity. The general theory states that space and time are variable and time slows down when in a strong gravitational field and the special theory states that time slows down when approaching the speed of light.

Both theories present time as being elastic, or relative. Now this word 'relativity' is the crucial clue to our understanding of the universe, or at least, our present understanding of the universe. Everything is relative to the state of consciousness. There are some scientists who are unhappy with the word 'relativity' because it doesn't sound objective enough and because objectivity is held as sacrosanct in science, but quantum physics has shown that objectivity itself is relative because the observer affects the particles' existence. Einstein is said to have considered calling his theory 'invariance theory'. Invariant means 'function remaining unchanged when specified transformation is applied', *Concise Oxford Dictionary*. The function remaining unchanged is the speed of physical light. Other things can alter around it such as time, space and mass but the speed of physical light remains fixed, more or less (with slight deviations). Everything else is relative to it, so 'relativity' theory is a very good name for it. If, as will be explained later, there are speeds faster than the speed of light, speeds of invisible light, then there would be relativity to those speeds also. So 'relativity' theory is perhaps the best name for it as it would still be a 'relativity' theory even if the laws were extended. The theory presented here is not to show Einstein's limitation but rather to extend the theory to embrace the anomalies in science, such as dark matter, wave-particle duality of light, gravity, the double-slit experiment and the 'spiritual phenomena' category of the anomalies in science, such as telepathy, near-death experiences, healing, remote viewing, etc. Most scientists' work is an extension of previous scientists' work. Einstein himself extended the theory of gravity which was originated by Isaac Newton who, previously, had expanded upon the work of Galileo Galilei. Einstein's work was also based on the field theory developed by James Clerk Maxwell, who had in turn expanded upon the work of Michael Faraday (1791-1867). It really needs to be grasped that scientific laws must evolve as new experience and observations become evident. As Einstein himself said "science is no more than a refinement of everyday thinking".

The problem for science today is that there are many anomalies which cannot be answered based on the presently accepted laws, such as the unifying of the four main forces in physics, the wave-particle duality of light, the extra gravity in the universe and also the conflicting theories of relativity and quantum theory. Both have been tested and are accepted as true theories of how the world works, but they are, or seem to be, contradictory. So a new theory needs to be found which not only reconciles both the relativity and quantum theories but that describes a whole unified theory of physics and which also takes into account the so-called 'spiritual' or higher dimensional realities of which the zero point field is included. These have to be included in a complete theory because this is where the extra gravity or dark matter resides and is the source of the, so-called, spiritual phenomena, the evidence for which is overwhelming   Spiritual phenomena is an integral part of human experience the world over so in order to have a complete unified theory this cannot be left out. To ignore such data would be very unscientific and, I would assert, a great loss, as answers to the anomalies in science can be found there. This theory then is what I call the Morphic Resonance of Light. Inherent in it is a new law which I am calling the Law of Resonance.

CHAPTER 8

A DIALECTIC THEORY

> *"The highest wisdom has but one science, the science of the whole, the science explaining the whole of creation and man's place in it."*
> Count Leonid Tolstoy, Philosopher and Author (1828 – 1910)

Relativity (sometimes called determinism because in it the future is already determined) describes the large scale. It's the Divine blueprint or master plan if you like because past, present and future all exist simultaneously. This means that the future is already determined. It has already happened, in other words. Relativity tells us that there is no point in making any choices because the future is already there. It's already known. So, in effect, we have no free will.

Quantum theory states that not only is the future not fixed but even each subatomic particle is unpredictable and we force it to make a choice as to where and when it will manifest by observing it. This is the same as saying that thought creates reality. It is saying that we have free will and it is by making choices that we choose our reality.

Both are well tested and accepted theories but are, apparently, saying the opposite to each other. They seem to be polar opposites, but this is the case with all dualism: – male – female, light – dark, hot – cold, implosion – explosion, centrifugal – centripetal, relativity theory – quantum theory, etc.. However, when we see them from a dialectic view they are seen to be reciprocal constants or complementary pairs which can be seen to be merged in a higher truth that comprehends them *(as described by Hegel)*. Relativity and quantum physics have a point of merging because both are speaking of space and time being variable or relative. **It's here that relativity and quantum physics merge. Both are speaking of relativity. Both are saying that time and space are variable and relative. They are relative to the speed of light which determines the state of consciousness.**

Einstein showed that time is an illusion and this has been proven in experiments. Time slows down at near the speed of light. Everything in our world is relative to the known speed of light. At speeds greater than the speed of light IT IS LOGICAL that time slows down further until eventually it stops and then there is no time. Harmonic multiples of the speed of light, each multiple speed having slower and slower time until eventually there is no time. This state of no time means that past, present and future do not exist as such; that they are all simulataneously occurring. This is relativity (or determinism).

Quantum theory says we are continually creating by our choices, so this implies there must be time, but I have already discussed how time is illusionary and that relativity says there is a state of no time. This contradiction is merely a limitation of intellect. The intellect is an aspect of consciousness but a limited aspect as it functions through humans. Through the evolutionary process a greater degree of intellect can manifest (because the consciousness is evolving) but the fact that we feel the contradiction means the intellect, hence consciousness, is limited to the speed of the third dimension. It sees everything through duality consciousness So relativity and quantum theory are another example of duality thinking due to the speed of our consciousness in the third dimension, but, they merge due to relativity of the speed of light which determines the state of consciousness. Another way of saying it is, both theories are true relative to the speed of consciousness.

So as not to upset quantum physicists, this does not preclude free will! All possible choices, multi-verses, parallel universes, potential outcomes etc. could be included within the no time state of relativity. It might be helpful here to make reference to cultural descriptions of this state. The Australian aboriginal culture and other indigenous people call this reality the 'Dreamtime'. They believe the universe was dreamed into being. Max Planck described the universe as a great Cosmic Mind. The physicist Sir James Jeans (1877-1946), described the universe as more like a great thought than a great machine. Einstein said 'the field' is the only reality. All these references point to the universe resembling a dream-like, mind stuff. It's tempting as humans, to place human limitations on a creative force but if we can stop doing this we can realise that a Cosmic Mind can include infinite possibilities. It could be that we have limited free will, but even if we have full free will the Cosmic Mind will encompass all of it because we are part of that Cosmic Mind, which includes all possibilities. We are the dreamer having the dream. !

CHAPTER 9

THE SOUL IS A HOLOGRAM

"As Above, So Below"
Hermes Trismegistus, Philosopher and Writer (pre - B.C. era)

So the speed of our consciousness determines whether we see the whole picture or stay locked in dualism. Dualistic thinking would say that relativity, being timeless, is cosmic will and quantum theory, which includes time, is human will. If it is true that we are made in the image of the Creator we must have a measure of will, and we do have will because this is the finding of quantum physics. We must be co-creators, but limited by the speed of consciousness. As above, so below is a spiritual maxim that was used by, and some believe to have been coined by, Hermes Trismegistus (in some cultures known as Thoth), an enlightened being, an avatar from the ancient Greek period who taught the holistic concept of the microcosm being a reflection of the macrocosm. This is the hologram concept. Each tiny piece of the image contains within it the information of the whole image. Being made in the image of the Creator is, I believe, referring not to the physical body image, rather the energetic image with all the same qualities in a state of potentiality – a hologram. What some may call the Soul.

Viktor Schauberger spoke of will and spirit as the principal causative forces of physical existence. This will and spirit are manifestations of consciousness. Or to put it another way, it is the method by which consciousness creates. How? Will and spirit are formative energies. When we talk of will or spirit they have to come from someone or something, a 'living entity' – human, sub-human or divine, but necessarily a 'living entity' or a consciousness in other words. Thoughts are the medium of conveying will and spirit from the living entity and it is thought which produces the form, because thoughts are lightwaves of consciousness and lightwaves have their own sound, so it's kind of like a stepping down process, or a slowing down process of lightwaves (consciousness becoming thought becoming sound becoming form). Another way of saying this is lightwaves/consciousness via thought/sound/OM produces vortex particles/forms. We usually regard thought as consciousness and use the words interchangeably but thought is only as aspect of consciousness and can be either constructive or destructive. Consciousness itself is much more fundamental than thought. Telepathy is a common experience and it tells us that thoughts do create sound. It's very common between mothers and their children. Sometimes it is just a sense impression but it is also possible to hear someone's actual thoughts though this is less common. The thought from consciousness creates the sound which creates the structure. This is the same as the statement made in Chapter 1 that Light (consciousness) is the breath of Source and the sound of the breath, Om, creates form. It is a fact that sound creates form and proof is presented in chapter 13 The 'Morphic Resonance of Light'.

So will and spirit come from consciousness and it is essentially consciousness which creates everything, from the galaxies to the subatomic particle. As everything arose from the Big Bang, then consciousness must be non-local, in other words, everywhere at once and there are many experiments which have proved this, some of which are described in later chapters. Consciousness is energy and as we know, energy and matter are the same thing relative to the speed of light. So at speeds greater than the known speed of light, there is greater consciousness. **It also means that matter is consciousness in form.** Energy becomes matter and matter becomes energy. The famous $E = mc^2$. This equation also describes what is happening in the quantum world. Matter coming in and out of existence. This equation also describes the whole life process, growth and decay, living and dying. Science tells us that all energy in the universe is conserved, it can never be created or destroyed. But this cannot be true if the universe is expanding. There is also the little matter of survival of consciousness after death and the subject of reincarnation, both of which are documented later. To understand how this affects the conservation of energy law consider that consciousness is energy. Energy is the ability to do work. Without consciousness we can do nothing, so consciousness is energy. Science insists that energy is always conserved in a dynamic system and that it can neither be created nor destroyed. So this means that when the body dies and decomposes into its natural elements, the consciousness which gave it the ability to do work, must be conserved. This is perhaps the strongest argument for survival after death but there is also ample evidence of this fact in the many studies conducted which are reported in Chapter 11 – Consciousness is the Key, and Chapter 21 - Reincarnation. **So the surviving consciousness may be what we term the soul. A spark of a greater flame or a small piece with the same potentiality as the larger piece; a hologram in other words. Co-creators limited only by the speed of our consciousness.**

As consciousness is an energy then it must be included in the E part of the equation $E = mc^2$. It is believed that Friedrich Hasenhorl, a victim of the first World War, was the first to put together this equation in 1903. Einstein saw it was the perfect equation to describe his ideas but this equation, like the known scientific laws today, only deal with the physical dimension, and as will be explained, this makes up only 5% (approximately) of Reality. Science already knows that approximately 95% of the universe consists of what they call 'dark matter' and they don't know what it is. The above equation can be adapted slightly to explain the other 95% of Reality of which the world is largely 'in the dark' *excuse the pun*. This will be explained in the chapter on dark matter. The answers to all the anomalies and unexplained mysteries to science lie within this 95% of Reality. It is beyond physical. So from the above equation we know that energy/consciousness and matter are equivalents but how does energy/consciousness become particles of matter? The answer lies in the vortex.

CHAPTER 10

VORTEX RESONANCE

> *The very word 'Universe' signifies a single curve (uni=one, versum=curve). The fact that the configuration of this curve may be a complex combination of descending and ascending, involuting and convoluting, expanding and contracting spiral movements does nothing to detract from its uniqueness or unit quality, since from inception to culmination its path is continuous. This curve is an energy-path and the essence of energy is ceaseless movement. In its eternal trajectory from spirit to matter (outward breath) and from matter to spirit (inward breath) it permeates all creation. It is all creation! "*
> Callum Coats, Scientist and Architect (Extract from 'Living Energies')

The vortex theory was popular in the 19th century, dominating physics and being taught at such august institutions as Cambridge right up until 1910. Indeed, the ether theory was taken for granted by physicists of that time and Lord Kelvin (1824-1907), the father of thermodynamics, believed that the whole universe consisted of two fundamental forms of motion: waves and vortices. The waves being light waves and the vortices being atoms, both arising from and existing in the ether. He believed that the vortex (spinning) nature of particles explained their mass, their elasticity and particularly their inertia because spin creates inertia. I have to thank the physicist, David Ash, (*The New Science of the Spirit*) for introducing me to vortex theory. He, in turn, learned of it from ancient yogic philosophy. So this is not a new idea and it is a great pity it was lost from scientific reasoning in the 20th century because it explains so much.

The <u>key</u> to understanding all of the 100% of Reality lies in the understanding of how consciousness, via sound, produces matter. It is done through a process known as vortex resonance. This means that every galaxy, planet, piece of dirt, atom, right down to every sub-atomic particle is a vortex, in other words, <u>spiralling</u> energy (consciousness), spinning in on its own axis producing the sub-atomic particles and at the same time resonating (vibrating) to a particular frequency from the morphic resonance field. So each particle has rotation and vibration or spin and frequency and each particle takes it place within the pattern created by the morphic resonance field to create shapes and forms. The spiral can be found throughout nature, from the sub-atomic world, e.g. when neutrinos are born they always spiral away in a left-handed direction, up to the everyday world where we see the spiral in the form of DNA, pine cones, rams horns, sea shells, etc. and on up to the spiralling galaxies. Even water naturally spirals. It always follows the path of least resistance, which is a spiralling motion. If you let go of the end of a water hose, it snakes around and around because the water inside is following the spiralling motion. Any natural watercourse will also create a spiralling pathway rather than flow in a straight line because it follows the path of least resistance and if you observe water vapour rising from a hot bath it swirls around in spirals as it rises up. So the spiralling motion is a fundamental principle. A vortex is spiralling motion on one point, or spinning in other words. **So matter is composed of spiralling energy which is spinning on one point, creating the sub-atomic particles, then atoms, then planets, then galaxies, each an increasing size of vortex.**

Atomic theory told us that each subatomic particle was a particle <u>in</u> motion but superceding this is the quantum theory which tells us that each subatomic particle is a particle <u>of</u> motion. In other words the particle is simply energy in motion, or consciousness in motion. It is not an actual solid thing. In 1928, the English physicist Paul Dirac (1902-1984) discovered that electrons exhibit a spinning motion and he also went on to discover the anti-electron, known as a positron (a positively charged particle). When the electron and positron meet they annihilate in a flash of light. This is because it is light that they are made of. Each particle of matter is simply spinning light, which makes matter all rather ephemeral. Particles both absorb and emit light, making them dynamic little vortices. The dynamism of light continually being absorbed and emitted allows a potentiality factor into the equation as each particle is neither static nor fixed but dances its piece to the particular tune as long as the tune is playing. Potentiality is therefore inherent in all subatomic particles, hence matter. The potentiality of light that is, to manifest from any of its many frequencies of dimensions.

Viktor Schauberger understood this potentiality factor of physical existence in which he believed *will* and *spirit* to be the principal causative forces, as described in the previous chapter. In 'Living Energies' which is an exposition of concepts related to Schauberger's theories, Callum Coats describes how Schauberger viewed will and spirit ..... "They deploy themselves through the agency of various lower orders and magnitudes of energy belonging to the 4th and 5th dimensions, i.e. through those more subtle, non-spatial dimensions of being that are inherent, but are not perceived in the three dimensional world to which we are accustomed. Of ethereal nature and endowed with very high frequencies and formative potencies, they could also be termed 'potentialities', which in their extremely sensitive and unstable state of energetic equilibrium await the right stimulus and occasion to manifest themselves". [1]. (This is quantum physics). The stimulus and occasion to manifest would be the will of the observer (in other words thoughts which arise out of consciousness). So I believe that the missing factor which is needed to reconcile relativity and quantum physics is consciousness itself As explained earlier, they merge due to relativity, of the speed of light which determines the state of consciousness. If your consciousness is operating at top speed you experience a state of no time, if it is operating at a lower speed (say 300,000 km per second, our known speed of light) you experience time.

The speed of light determines the state of consciousness because each harmonic multiple of the speed of light determines the frequency of light waves, each multiple having ever ascending frequencies. The speed of light is part of twice the speed of light, both of which are part of four times the speed of light, etc. If you remember in the Introduction it was explained that each octave is double the frequency of the previous. So within the cosmic harmony, each ascending octave is double the frequency, x2, x4, x8, x16 etc. This enables a phase relationship of waves, which means they embed within each other, and this creates coherence. Coherence, (meaning embedded together, focussed, not scattered), therefore, creates accumulation of information, so whatever is in the lower speed must be in the greater but in greater quantities because the lower is a subset. The level of consciousness in the lower speed is a subset of the level of consciousness in the greater speed, so at greater speeds all information is known.

Greater speeds mean greater consciousness.

CHAPTER 11

CONSCIOUSNESS IS THE KEY

> *"The universe appears to have been designed by a pure mathematician and it begins to look more like a great thought than like a great machine"*
> Sir James Jeans, Physicist (1877-1946)

Consciousness is a rather taboo subject in scientific circles, generally speaking, because it is a 'mystery' and can't be explained conventionally. It has tended to be glossed over in any scientific discussion on physics, biology or chemistry, although there are some physicists now who are beginning to realise that consciousness has a prime role to play in understanding physics. Some neuroscientists describe it as a function of the brain but new studies are showing that the brain, rather than being the producer of consciousness, is simply a receiver and that consciousness exists in every cell. Also, that individual consciousness can exist outside the body and can travel, such as happens in near-death experiences (NDEs).

There has been much anecdotal evidence of near death experiences and a comprehensive collection of these was published in the 1970's in a book by Dr. Raymond Moody, called *'Life After Life'*. However, as we know science requires proof. This is beginning to emerge now in several studies. Some of these are being conducted by a former student of Dr Moody, Dr. Bruce Greyson, who is Professor of Psychiatry and Director of the Division of Perceptual Studies at the University of Virginia and a Committee member of the International Association of Near-Death Studies (IANDS). [1]. IANDS, in existence for almost thirty years, is an important network for researchers and academics in this field, as well as offering support to NDE experiencers and their families. What Bruce Greyson and his team have found is that in most instances the person returns from the NDE with a profound change of attitude often resulting in a change of career. Their attitude changes to a less materialist view of life towards a more spiritual and compassionate approach. He cites a drug criminal who becomes a counsellor for deprived children and a police officer who found he could no longer shoot a gun, becoming a high school teacher.

Another study is being conducted by Dr. Peter Fenwick, a consultant neuropsychiatrist at the Institute of Psychiatry in London, Dr. Sam Parnia, a clinical research fellow and registrar at Southampton General Hospital and Dr. Pim van Lommel, a cardiologist from Arnhem, Holland. [2]. Although NDEs have been reported for centuries, they have usually been treated with scepticism. The early findings of this new study, however, show that consciousness can exist outside the body. Some neuroscientists remain very sceptical and believe that the experiences people are having while they believe they are dead are actually happening either at the point of losing consciousness or at the point of regaining consciousness. However, this argument falls flat when studying some cases. Some patients have correctly recalled conversations between the medical staff and accurately described instruments and procedures used on them *while they were dead and being clinically monitored.* This has been verified by medical staff

Perhaps the most convincing case was reported in a BBC Horizon programme in February 2003. Pam Reynolds from Atlanta, Georgia underwent surgery to remove a deep aneurysm in the brain. Dr. Robert Spetzler, a neurosurgeon from Phoenix, Arizona carried out the operation against all odds. She had been told by her neurologist that she had zero chance of survival. Her body was cooled to 15^oC and all blood drained from her brain. Her brain was devoid of activity for at least 1 hour. During that time she left her body and from a vantage point above her body, witnessed the procedure being conducted on her body and also conversations which took place in the theatre. She then travelled through a tunnel of light, met deceased relatives and others and she reported that everything was light and felt a deep sense of peace, consistent with other experiencers of NDEs. She asked if the light was God and was told that the light is what happens when God breathes. She reluctantly returned to her body at that point with encouragement from a (deceased) relative but afterwards, her recall of all that had transpired in the operating theatre, including accurate description of instruments used and conversations between staff was verified by the medical staff, who have no explanation, since everything she reported happened while she was in deep anaesthesia when there was no brain activity and she could not have seen or heard anything. What this study shows above all else is that consciousness is not confined to the brain, or body. It can travel and has an independent existence of the body. The body and brain are instruments for it to manifest through.

This seems to be the conclusion reached in another study being conducted by the British Physicist, Roger Penrose and Dr. Stuart Hameroff, Anaesthetist and Director of Consciousness Studies at the University of Arizona. Dr. Hameroff feels that understanding anaesthesia is a clue to understanding consciousness. [3]. They believe that consciousness exists as wave patterns in the smallest spaces of atoms (remembering that atoms are mostly space) and during anaesthesia or death, the waves leak from the body and somehow coalesce creating the conscious state outwith the body. However, this doesn't describe the tunnel and all the experiences within that.

Many patients report being aware of looking down on their body, often from the vantage point of the ceiling. They often become aware of a point of light which becomes a tunnel and they feel pulled towards it. They also say that as soon as they think of being somewhere they can travel there instantaneously and walls, distances etc. are no barrier. They also, almost unanimously, report seeing a pinpoint of light which then becomes a tunnel and they feel drawn to it. On entering the tunnel they see people known to them including deceased relatives, friends and others unknown to them but that they feel are somehow connected to them. The interesting thing about this is that the people all seem to be made of light. They also sometimes report seeing what they consider to be heavenly or divine beings and in most cases, report feelings of deep peace and overwhelming love, so much so that they are often reluctant to return to their body when told they have to (usually because they have unfinished work on the Earth plane). They also sometimes report receiving guidance of some sort and this is often what creates the change of attitude and career.

With the ongoing studies a picture is emerging that consciousness exists in every part of the body and outside the body, and can be anywhere instantly with intention. This picture is consistent with the description of non-locality and is further proof that consciousness is light because light is non-local, and the fact that everyone seems to be made of light and the description of the tunnel of light is consistent with the description of higher harmonic light existing beyond the physical dimension and the vortex principle. The vortex being the tunnel, as a vortex naturally 'pulls' into the centre. These studies seem to suggest that the world outside of the physical body is a world of light.

This light is the lightwaves of consciousness which, as already described, is non-local so that is why it can be everywhere at once and can travel anywhere instantaneously with intention. This describes what is happening with remote viewing, clairvoyance, distant healing, prayer etc. With remote viewing, for example, it is possible to travel anywhere and see anything while in the comfort of your armchair because consciousness is not confined to the body. It can be anywhere, with focussed intention, and is already everywhere anyway. Remote viewing was practiced by both American and Russian governments as spying tactics against each other in the cold war, with remarkable success, and probably still continues to this day. Potentially anyone has the ability to do this and it is possible now to attend training courses to learn this skill. Experiments into remote intention conducted by PEAR labs at Princeton, New Jersey, have been very convincing and these are discussed below.

So it is no longer sensible to believe that consciousness is a function of the brain and ends with death of the body. It is true that the brain is an instrument through which the consciousness reveals itself, but its existence is not dependent on the brain.

Consciousness is the primary source of Being from which OM sounds forth. There only *is* consciousness. It breathes out OM when it wants to create Form.

Consciousness as Relativity is the Omniscient Consciousness, the Plan, the Blueprint. - At Source - Reality Is. It is simply pure Beingness - without time, so all past, present and future must co-exist.

Consciousness as Quantum physics is when it breathes out OM through an act of will and creates many forms of itself. These forms (a human for example) use will (because we are the consciousness we also have creative will) to create forms also. Quantum physics, remember, tells us that it is through observation (act of will) that we bring particles into existence. It is the act of will (through observation in experiments and through beliefs in our daily lives) that we create our perceived reality.

This act of creating and changing physical matter with our will/consciousness has been verified by Brenda Dunne and Robert Jahn of PEAR Lab at Princeton University in New Jersey. PEAR stands for Princeton Engineering Anomalies Research and was set up to investigate the influence on mind over matter at a distance and other such anomalies. They have amassed perhaps the largest collection of data anywhere on such experiments. On studies into remote intention using random event generators (REGs) a statistically significant outcome was achieved which the US National Research Council concluded could not be explained by chance. [4]. The kind of numbers we are dealing with here are approximately 2.5 million trials over a 12 year period where nearly two thirds of participants were able to influence the machines according to their intentions. [5].

Science can tell us that matter is a sort of frozen energy, but not what energy actually is. They can measure its effects and use it in equations but as the physicist, Richard Feynman said, "It is important to know that in science today we do not know what energy is". It is defined as 'the ability to do work'. As we have seen earlier that consciousness influences the random event generators at PEAR labs, this proves that consciousness is energy because it creates the ability to do work. This also means that consciousness must survive death of the body, as discussed earlier, because as we know, all energy is conserved. Without consciousness we can do nothing. Consciousness has been described earlier as spiralling light waves, the zero point field or the dark matter. As this dark matter is flowing unheeded throughout cosmos in unfathomable amounts, then the energy described in $E = mc^2$ is an unfathomable energy, which means that the mass must be equivalent to it. There is obviously not enough mass in the universe, as presently understood, to equate with that much energy, so the second part of the equation needs to change. It needs to change to reflect the finer matter which exists in the higher dimensions. This would explain where the extra gravity is coming from. Perhaps a more inclusive equation to describe the universe would be-

Consciousness = vortices x speed of light $^{\text{harmonic multiples}}$

Or to adapt Einstein's equation - $E = mc^n$ (n being an unlimited harmonic multiple)

n becomes a cosmological variable and would account for the missing mass in the universe and would explain time differentials. It is a description of the harmonic multiples of the speed of light, or higher dimensional light.

So with this model of consciousness we can see that not only does consciousness survive death of the body but that it is omnipresent. Since there are harmonic dimensions of light, this means there are harmonic dimensions of consciousness – it's the same thing. In esoteric philosophy, the Ageless Wisdom teachings for example, state that the person's level of consciousness at the point of death determines where the consciousness goes to, which dimension in other words. This makes logical sense when you consider what was said earlier about consciousness being determined by the speed of light. When a person has quickened their consciousness in life it will literally be operating at a faster speed, so, on being released from the body it would naturally exist in the relevant dimension based on its speed. The soul which inhabits the physical body is naturally operating at a faster speed anyway, which is why we can't see it physically. At death, it leaves the body, often through the top of the head, sometimes the heart area, and many medical professionals have witnessed what looks like a blue electric light leaving a body at the point of death. Many souls go to the astral or etheric planes, as they are called, which are operating faster than the physical dimension but there are many dimensions beyond them. These two planes are relatively easy to contact by mediums and many bereaved people have been helped enormously by receiving irrefutable proof that their loved ones still exist.

It is important to realise that just because a person has 'died' and now exists in a different dimension does not mean that they are enlightened or advanced. I mention this only because a gullible person may believe that because a message is coming from a different dimension that it must necessarily be profound wisdom. There are some mischievous entities existing on the lower astral planes who can easily channel through mediums and present themselves as 'advanced teachers'. Discrimination must be exercised and the quality of the information will be the giveaway. The higher, or faster, dimensions are harder to contact unless sufficient inner work has taken place, to attune the consciousness and open the heart centre. As will be explained in future chapters, the opening of the heart centre is crucial to the evolvement of consciousness. So we can see that there are many levels of consciousness, or dimensions if you like. Consciousness *implies* intelligence, which means that the entire cosmos is an intelligence.

This realisation is increasingly reflected in the thoughts of some astronomers, astrophysicists and biologists who believe that the universe seems to have been finely tuned to enable life to exist; that it could not have arisen by 'chance'. It seems that it has been intelligently designed. Even Max Planck, one of the founding fathers of quantum theory believed that a conscious intelligence was the precursor to all matter, as quoted in a lecture given in Florence, Italy, he said "As a man who has devoted his whole life to the most clear-headed science, to the study of matter, I can tell you as the result of my research about the atoms, this much: There is no matter as such! All matter originates and exists only by virtue of a force which brings the particles of an atom to vibration and holds this most minute solar system of the atom together …. We must assume behind this force the existence of a conscious and intelligent Mind. This Mind is the matrix of all matter." This conscious intelligent Mind is the central theme of this theory; that matter and space (which includes dark matter) is all spiralling lightwaves of consciousness and the 'force' which he speaks of which is holding all matter in place, is the morphic resonance fields, or sound, or OM, however one wants to describe it. We can now see that the E for energy in $E = mc^n$ is descriptive of consciousness, so it is perhaps more true to describe this equation as $C = mc^n$. (C for consciousness).

CHAPTER 12

THE BLASPHEMY, or THE BEHAVIOUR OF LIGHT

"The changing of bodies into light, and light into bodies, is very comfortable to the course of Nature, which seems delighted with transmutations."
Sir Isaac Newton, Physicist and Mathematician (1642 – 1727)

For an explanation of the actual physics of this we now have to turn to light - the wave-particle duality of light (another anomaly for science and another example of duality consciousness). Consciousness expresses in light waves, physical and non-physical. Science can measure the physical waves, but light waves exist in all dimensions beyond the speed of physical light. However, the blasphemy doesn't end there. I have mentioned previously that lightwaves exhibit spiralling motion. The spiral being the most basic form throughout the universe and found throughout all of nature. In physics we are told that light, as well as being a particle (a photon) is also a wave. The normal way of thinking of this wave is in a two-dimensional image. However, if we think of an electrical wave, we know that as it travels it creates a transverse wave. Which means that the electric and magnetic fields vary in a periodic way at right angles to each other and to the direction of propagation.

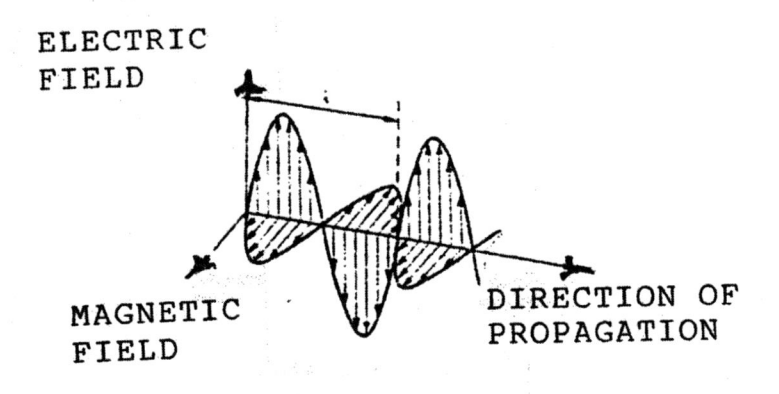

```
     ELECTRIC
     FIELD


     MAGNETIC              DIRECTION OF
     FIELD                 PROPAGATION
```

FIGURE 4 - TRANSVERSE WAVES OF ELECTROMAGNETISM

So we can see that the electromagnetic wave is three-dimensional. Since everything is made of electromagnetic waves or frequencies and it has been argued that everything is made of lightwaves, (meaning the electromagnetic waves are lightwaves), then lightwaves must be three-dimensional. This has been found to be the case in experimentation, described below but it is also consistent with the whole theory presented here, that lightwaves are spirals (spirals being three-dimensional). This means that lightspeed must vary according to frequency because the speed would be determined by the angular acceleration and radius of the spiral The speed determined by the angular acceleration and radius of the spiral was the view of Walter Schauberger and evidence of this spiral movement of light was produced by Prof. Felix Ehrenhaft at the Physics Department of Vienna University in 1949 [1]. The following explanation is taken from *'Living Energies'* (p24) by Callum Coats. Prof. Ehrenhaft performed

"...... experiments with barely perceptible particles of matter and gas particles enclosed in glass tubes which were observed when illuminated by concentrated light rays of various frequencies. Observations of this phenomena were made under conditions varying from high pressure to high vacuum (30 atm to 1×10^{-6} mm Hg (Hg = mercury) and it was concluded that since the spiral movement of the observed particles was caused by light rays, the particles had to be propelled along the same spiral path as the light itself." [2]. He called this process 'photophoresis'.

The evidence that particles are little vortices of spinning light arises from the point above; **spiralling lightwaves held on one point would naturally spin**. Also, as described previously, the vortex principle follows the fractal principle. So whatever is in the larger will be replicated in the smaller; spinning galaxies and planets will be replicated in spinning atoms and sub-atomic particles and they are produced in the same way, from spiralling lightwaves held on one point. Also, we know from experiments with gamma rays that sub-atomic particles can be produced from light. We know that when matter and anti-matter meet, such as electrons and positrons (opposite charges) they annihilate in a flash of light. We know that atoms emit and absorb light so it is accepted that light is transformed into matter. This is $E = mc^2$ or $C = mc^n$. So we can see that energy, or consciousness, is light. In splitting the atom, nuclear radiation is released but nuclear radiation is radiatory light (at very high frequencies). It is also accepted that light cannot be brought to rest. It has no rest energy, therefore the particles (made from light) must be in constant motion or perpetual motion, as we know they are and the motion is that of spin. So we have a consistent explanation here of lightwaves being spirals and particles are the arrested spiralling lightwaves, which creates spinning. Since particles absorb and emit light, they are dynamic. This dynamism is the perpetual motion and it will continue as long as the resonance is held. In other words, as long as the sound exists because it is the sound which is creating the little vortex. **So we can say that**

sound is the cause of the perpetual motion of the particles of matter.

In physics, light can be both a particle, which is called a photon, and at the same time behave like a wave. This is the anomalous wave-particle duality of light. For example, if you shine light through a pane of glass, only about 95% will be transmitted through it, the remaining percentage, about 5% will be reflected back at you and this is why you see your image, faintly, in the glass. Now the problem is, photons, which are identical little particles of light, when encountering the glass should all behave in exactly the same manner because they are identical. So they should all either transmit through the glass or all be reflected back, but they don't. 95% go through, 5% reflect back. This is the behaviour of waves. So particles of light behave like waves.

This phenomenon was discovered in the 1920's and physicists decided that this must mean that photons, although identical, were displaying properties of chance or probability. They all had a probability of being transmitted at 95% or being reflected at 5%. They also decided that this must apply to all other sub-atomic particles also which meant that everything in the universe operated on chance. This outcome was deeply unsettling to Einstein as it contradicted his theory of relativity which says that the future is already determined (Special Theory) and that the movement of planets and other bodies is predictable (General Theory). This is when he made his famous statement that 'God does not play dice with the universe'.

Now I think what is happening here is that in respect to light behaving as a wave, light is in essence a spiralling wave anyway, and as described in the chapter on Vortex Resonance, it becomes a particle due to resonance. So it becomes a particle, or photon, **but even when it is a particle it still displays wave behaviour because waves or periodicity and their frequency is an inherent pattern in all of nature. It is inherent in the Morphic Resonance.** More of this will be explained in the next chapter. So what we see in the pane of glass example is periodicity or frequency in action. Periodicity or frequency is a fundamental function of nature and is inherent in the morphic resonance fields of all things because morphic resonance means 'the creation of form through the re-sounding of wave frequencies'.

There is no logical reason why identical particles would display a ratio of 95 : 5 chance of behaviour without an underlying cause for it. Morphic resonance explains this. The morphic resonance field of the pane of glass is determining the behaviour of the photons. Like in the famous double slit experiment taught to all undergraduate physics students, the photons have no choice but to take their place in the existing pattern of the morphic field (explained more fully in Chapter 14).

To make a slight digression here, something needs to be said about the word 'chance'. We commonly take it to mean, random, unpredictable or unknown. However, when we look at divination this changes our perception of that word. Divination is often described using words such as random or chance, such as picking a card at random, for a card reading, or the Chinese practice of throwing the coins at random, etc. Divination can take many forms and is practiced throughout the world by many cultures. Coins, cards, sticks, in fact almost anything can be used, even tea leaves, to see the future or gain an answer to something, by throwing the coins, sticks, or picking the cards etc. at random.

Now much fear and distrust surrounds these practices particularly by some religious groups because they do not understand the process. What is happening is that the person is apparently throwing the coins or sticks at random, but prior to this, they have asked a question and a higher force operating at a faster speed of consciousness where, remember there is greater consciousness and time is non-existent, creates the outcome to give the answer. It is not chance (as we commonly understand the word) which is operating at all. The higher force or greater consciousness, whichever you prefer to call it, is really an aspect of our own consciousness which is operating at a faster speed. If you remember the fractal principle, that our morphic field is embedded within the greater morphic fields operating at faster and faster speeds creating slower and slower time, eventually a state of no time where past, present and future are one and the same. When this consciousness is accessed, all answers can be found, all information is known because all time is happening at once. This is a description of Relativity. It is the same with dowsing. When a question is asked, the pendulum or sticks, give the answer by swinging one way or the other. The pendulum or sticks are nothing more than a tool to show, physically, what the answer is. They are being made to move by the person's own hand, directed by their own greater consciousness, or higher self as it is sometimes referred to. There is really no mystery to it. It is just tuning into a faster speed of consciousness. The tools used to display the answer are just that – tools, to show us physically what we need to see. Some people don't need tools at all, they just know the answer without having to see it physically. This is intuition. Dowsing is also known as divination and the word 'divination' means 'consulting the divine'. So we are consulting a greater, or more divine aspect of ourselves, which some may call the Soul.

So when we talk about chance happenings, it is probably just that we cannot see the bigger picture, or 'the plan'. We've all had the experience of thinking about someone then suddenly they call or we bump into them. Most people call this coincidence or chance but really it's just that, at a higher level, we know it is going to happen and we've temporarily tuned into that knowledge. We've temporarily tuned into a faster speed of consciousness. This is intuition. It functions in everyone without exception but is mostly dismissed or suppressed due to conditioning. Most people can recall an instance where they went against that inner knowing and listened to the logical, left brain voice instead and then found that things didn't turn out so well. Intuition is very often active in mothers and is most likely a protective mechanism. Everyone can develop their intuition with practice though if the left brain has been very dominant throughout life it may take a little more practice.

As far as photons and other sub-atomic particles are concerned, as already explained previously, I believe they exhibit potentiality but potential is not the same as probable or chance. In the double-slit experiment, the particles are not displaying chance (in the common understanding of the word), they are taking their place in the morphic resonance field; or the template. So when quantum physicists talk of the universe being based on probabilities or chance (in the common understanding of the word) I believe this is a great mistake. What it is based on is potentialities which obey certain laws, for example, periodicity. This means we are co-creators, creating with our will (quantum physics), but does not necessarily dismiss another co-creator which *contains* all the potentialities (relativity).

With probability on the other hand, there can be no reconciliation of quantum theory and relativity theory. Probability or chance, as we commonly understand it, is not logical or intelligent, nor does it display meaningfulness. As I have already alluded to previously, it is now a consensus among many scientists that there has been a fine-tuning of design in the cosmos which means there must be a designer; an architect; an intelligence. If you remember in Chapter 10 on Vortex Resonance, I referred to Viktor Schauberger's explanation of 'will' and 'spirit' creating from the potentialities of atoms. Will and spirit necessarily implies a consciousness, designer, architect or creator with unlimited potentiality. This is simple, logical and intelligent whereas probability is immature, unclear and hard to understand. If you remember Schauberger's quote at the beginning of Chapter 1 "the majority believes that what is hard to understand must be very profound. This is incorrect. What is hard to understand is immature, unclear and often false. The highest wisdom is simple and passes through the brain directly into the heart".

In the Introduction I explained how mainstream science is reluctant to peep outside the box of known laws and that they try to make new discoveries fit into the existing laws. Another example of this is non-locality. This means that when two particles are born together they are forever connected instantly no matter how far apart in distance they ultimately become. What affects one particle, affects the other instantly, in violation of the speed of light. As all matter originally arose from the Big Bang, this means that everything in existence is non-locally connected instantly throughout the universe and this is the central theme of this book, that light or consciousness is the sum total of everything and it is non-local. Speeds and time are simply a construct relative to the speed of consciousness. Non-locality was known about in Einstein's time and he believed, because of this that quantum theory had to be wrong, because it appears to violate the speed of light. But experiments have shown non-locality is a fact. Now, despite this, mainstream science insists that quantum theory is still compatible with the speed of light. In his book 'The Universe Next Door' the author Marcus Chown speaking on non-locality states "…only limited kinds of information can be transferred instantaneously between particles. It would not be possible, for instance, to send a meaningful message". What an extraordinary statement! All this statement shows is a limitation of vision. Let us just examine this statement "only limited kinds of information can be *transferred instantaneously* between particles". This is an admission that faster than light speeds do exist. Now the next question to be asked is "what is a meaningful message and who is to decide what is meaningful?" I have already argued that a particle of light is meaningful because it creates matter, in fact it *is* matter and, I have argued, it is a spinning wave of consciousness. Also, as I discuss in Chapter 15 Proof of Speeds Faster than Light, faster frequencies contain more information. This is because waves are carriers or encoders of information. That is why everything has a frequency 'signature'. The signature is the unique code of information for the particular item in question. When waves interact or interfere with each other, such as interference waves (as in the double slit experiment) they combine their information, so interference waves contain an accumulation of information. Therefore, when there are more waves (higher frequency) and more interaction between them this creates an accumulation of information. This can be shown visually in the photos of morphic resonance created by different frequencies. The higher frequencies display more complex patterns. So particles at very high frequency could contain potentially limitless information.

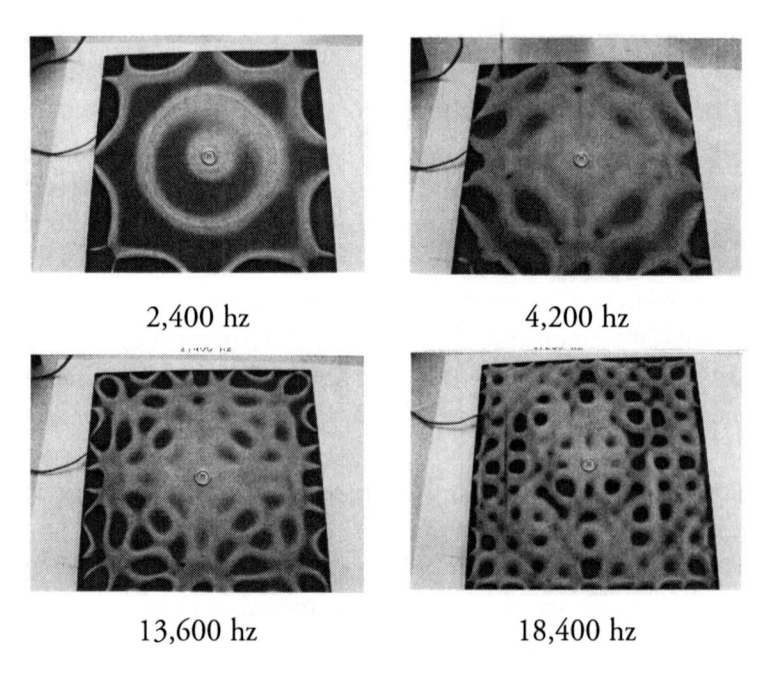

2,400 hz	4,200 hz
13,600 hz	18,400 hz

FIGURE 5 - MORPHIC RESONANCE PATTERNS CREATED BY ASCENDING FREQUENCIES

These experiments were conducted at Aberdeen University, Physics laboratory. The plates used were aluminium, 1.6mm width and 200mm x 200mm square. The plates were placed on top of a vibration generator which was attached to a frequency generator and amplifier. Fine sand was scattered loosely on the surface of the plate. The frequency generator was 'tuned' to various frequencies which produced the patterns from the sand.

It has been discussed how particles are made from spiralling lightwaves of consciousness spinning on one point and that this occurs due to sound or resonance, but the question to be asked is, what creates the pattern or template for the particles to produce form?

For spiralling lightwaves of consciousness to create form, the physical stuff of matter, the light has to be arrested onto one point to produce sub-atomic particles, the vortices as described earlier in Chapter 10. These then have to be arranged in patterns to produce atoms and then greater patterns to produce matter. These patterns are in existence in the OM, the primal sound, which are "...Of ethereal nature and endowed with very high frequencies and formative potencies" *Viktor Schauberger*. Sound does create geometrical patterns which can be proved easily by experiment, and explained in the next chapter on Morphic Resonance of Light. It is the sound, hence pattern, which forces the spiralling lightwave to spin in on itself becoming a particle which then takes its place within the geometrical pattern to create form. The geometrical pattern is held as long as the sound is held. This is known as resonance, which means resounding, echoing, continuing to sound, reinforced or prolonged by vibration or reflection. So the spiralling lightwave is forced to take the shape of the pattern caused by the resonance and the only way it can do this is if it becomes a particle, otherwise it would continually radiate away as a spiralling lightwave and not take form. This can be easily seen in morphic resonance experiments. For example, if a substance such as sand is scattered on a plate, then a particular frequency is sounded against the plate, immediately, the sand arranges itself into a pattern and it can be any kind of pattern. The higher the frequency, the more complex the pattern, as shown in Figure 5. **So sound compels the construction of form**.

Another way of saying it is, that the particle is resonating to the shape created by the sound. Another way of saying it is, that geometrical patterns freeze light waves into particles and so create form. Another way of saying it is, matter is suspended or spinning light in geometrical compositions created by resonance. Spinning is a fundamental principle throughout cosmos. All particles spin, as do all atoms, planets, galaxies - even a newly fertilised human egg begins to spin. Again we have the principle - as above, so below at work.

CHAPTER 13

MORPHIC RESONANCE OF LIGHT

> *"The more one studies these things, the more one realizes that sound is the creative principle. It must be regarded as primordial. No single phenomenal category can be claimed as the original principle. We cannot say, in the beginning was number, or in the beginning was symmetry, etc. These are categorical properties which are implicit in what brings forth and what is brought forth. By using them in description we approach the heart of the matter. They are not themselves the creative power. This power is inherent in tone and in sound."*

Hans Jenny, Medical Doctor, Philosopher and Scientist (1904 - 1972)

Sound may well be the instrument used to create the beautiful and increasingly complex crop circles and patterns appearing in fields around the world. They appear in hundreds of countries around the world, even in snow and ice in remote regions, but the bulk of them have appeared in south-west England around the Stonehenge/Avebury area. Who or what are creating them is the subject of another book but the patterns reflect the principles of sacred geometry which is, apart from looking beautiful, a language which speaks to the subconscious. The dedicated researchers who include pilots, farmers, architects, mathematicians, doctors, historians, healers and many others have discovered hidden information, for example calendrical systems, higher mathematics and evolved geometry (beyond Euclidean or anything seen before) within the formations. But this is the case with sacred geometry, each pattern or symbol is a mine of information. Symbols are the language of the subconscious, which is why we tend to dream in symbols. The great psychologist, Carl Gustav Jung (1875-1961) in his bestseller book *'Man and his Symbols'* explained his theory of the significance of symbolism in dreams and art and that the understanding and practical advice received from dreams can lead to greater self-understanding and a more productive life. The book was itself a result of one of his own dreams, the symbolism of which he decided to act upon. Symbols or patterns are a form of pictorial thought and enable us to find ways and means to see *behind* things, *behind* the obvious. It is a language with multi-levelled information and meaning.

One of the first people to discover that sound creates patterns was Ernest Chladni (1756-1827). He discovered that by stroking a metal plate covered in powder with a violin bow, different notes (frequencies) produced different patterns in the powder. The record of these experiments gave the world the experimental principles of acoustics in such works as *'die Akustik,'* in 1802. Following this, Swiss doctor and philosopher, Hans Jenny, did extensive research on the effects of sound to create form and a summary of his work is reported in a two volume book called *'Cymatics'* [1]. As a young boy he discovered the science of anthroposophy, or spiritual science, proposed by Rudolph Steiner (1861-1925) and Johann von Goethe (1749-1832), a major element of which incorporates sacred geometry, in other words, the significance of form and number and its implicitness in all of life. He experimented with various powders, pastes, sand etc. on a metal plate. By attaching an oscillator to the plate he was able to generate a wide range of frequencies onto the plate. Each different frequency produced a different pattern in the sand or paste etc. Whenever the frequency, or sound, changed, the pattern changed. He called this process Cymatics, after the Greek 'ta kymatika' meaning "pressure or waves" which is an accurate description of sound. Sound is nothing more than the frequency of pressure waves. What he discovered, as well as proving that sound creates form, is that the underlying principle of all of nature is periodicity or vibration which is the function of frequency. Frequency can be described as the number of waves in a given unit of time. **So all of life is based on the frequency of pressure waves, or sound.** We know from studies of the quantum world that everything is constantly vibrating. Even the Sun itself is in constant vibrational motion with many different complex oscillation patterns being displayed. This was discovered in 1976 by Henry Hill and colleagues at the Santa Catalina solar observatory in Tucson, Arizona. This principle of periodicity was known in ancient times. The Greek philosopher, Heraclitus coined the phrase "All is flow. All is in flux".

Cymatics experiments also proved that different flow patterns are created by different frequencies. The flow can change direction of spin due to a different frequency. Also patterns and forms can either be static or can undulate, dance, crawl, revolve, invert, any movement at all in fact. There is no limit to the form or movement of either particles or the patterns that they make. It is all controlled by sound. The atom is vibrating to a frequency to take its place in a pattern, but within it, the electron and proton are spinning in accordance to different frequencies. We know they emit different frequencies of light so they must be operating at different frequencies within the atom. This implies that there must be frequencies within frequencies. The electron and proton have opposite electrical charges. As the cymatics experiments have discovered, different frequencies create opposite directions of spin and it could be that the electron and proton's direction of spin are relative to the stability of the atom. This would make interesting research. **We know that spin creates charge and spinning charge creates magnetism so that means every electron and proton is a little electromagnetic particle. This is also further**

evidence that electromagnetic waves are most likely spirals, because a spiralling lightwave would create charge which in turn would create magnetism, hence the electromagnetic wave which produces the electromagnetic particle.

As long as the sound is held, the form stays the same. In this way we can see that sound is the instrument of evolutionary change with genetic changes occurring due to a change of sound. Orthodox science believes that genetic mutation is the driving force of evolution but I would say that the cause is fundamentally sound. Sound creates genetic changes. If the sound changes the form changes. This is morphic resonance. It is the resonance which allows the form to take shape, or another way of saying it is that the form grows into the shape created by the sound.

This resonance is the morphogenetic fields described by biologist, Rupert Sheldrake. Sheldrake predicts these fields are the reason why organisms grow into particular forms. Science, at present, doesn't understand why form takes the shape it does. It is not due to DNA, as commonly assumed, because as Sheldrake points out, in any flower e.g. the leaves, roots and petals all share the same DNA but their forms are different. So there has to be something else to explain the growth of form. I believe the geometrical pattern created by the resonance (from the OM) creates the growth of form. It is already in existence and the plant grows into the field or geometric pattern. Sheldrake also ascribes memory to these fields and uses a brilliant example to illustrate this. The example is of milk-drinking blue tits. In the early 1900's doorstep milk deliveries began in England. After a couple of years it was observed that in the town of Southampton, the local blue tit population began to peck at the bottle tops and drink the cream. Shortly thereafter it was observed that blue tits in other cities in England had developed the same habit. This began to be monitored by zoologists because blue tits are a home loving bird and tend not to stray far from their local territory. It was generally felt that it must be because they were making independent discoveries, 'a strange coincidence type of scenario', except that the habit spread and started to speed up. Sir Alistair Hardy, Professor of Zoology at

Oxford made the brave suggestion that it might be due to telepathy but wasn't taken too seriously. Even today the word 'telepathy' invites scorn by mainstream scientists but as has been discussed if everything is interconnected and originates from the same source, that is, the Big Bang, then all particles are non-local. Non-locality and proof from prayer experiments is described in Chapter 22 The Heart and Quantum Concepts and this would explain telepathy. **Non-locality enables telepathy to happen.** But back to the blue tits' story. It was discovered after a few years that blue tits in Holland were practicing the same habit. Where the story becomes really interesting is that with the outbreak of the second world war in 1939 the milk deliveries stopped and didn't restart again until 1948. Blue tits only live for 3 to 4 years and so none who were around when milk deliveries stopped would have been around in 1948 when they restarted. But in a short time the habit started again in the UK and Holland.

    This is a clear example of a memory field in non-physical existence. The blue tits learned something, which means they evolved into a higher, or faster, fractal of their morphic resonance field. The field does not just create the physical form (the blue tit's body) but has higher aspects to it which contain memory, feeling etc. (these would be the mental body, spiritual or causal body etc.). In each kingdom; mineral, plant, animal, human, there are fields within fields, or layers within layers, which is the fractal principle. There are ascending dimensions of harmonic multiples of the speed of light and each of the 'fields' exists in a different dimension, reflective of its frequency. These fields are what makes up the aura or life field around living things, including planets. Energy fields are known to be precursors to matter and can take the form of geometrical shapes. Also, each shape vibrates at different frequencies which means it would radiate different colours, the colour being dependent on the frequency. Many people with clairvoyance can actually see the different shapes exhibiting different colours. This is something else which was known by the ancients. In the "*Timaeus*" Plato describes how everything in the universe is composed of measure and number, proportion and mathematical form and that the basis of the structure of form is through certain shapes called polyhedra, which later came to be known as platonic solids (after Plato).

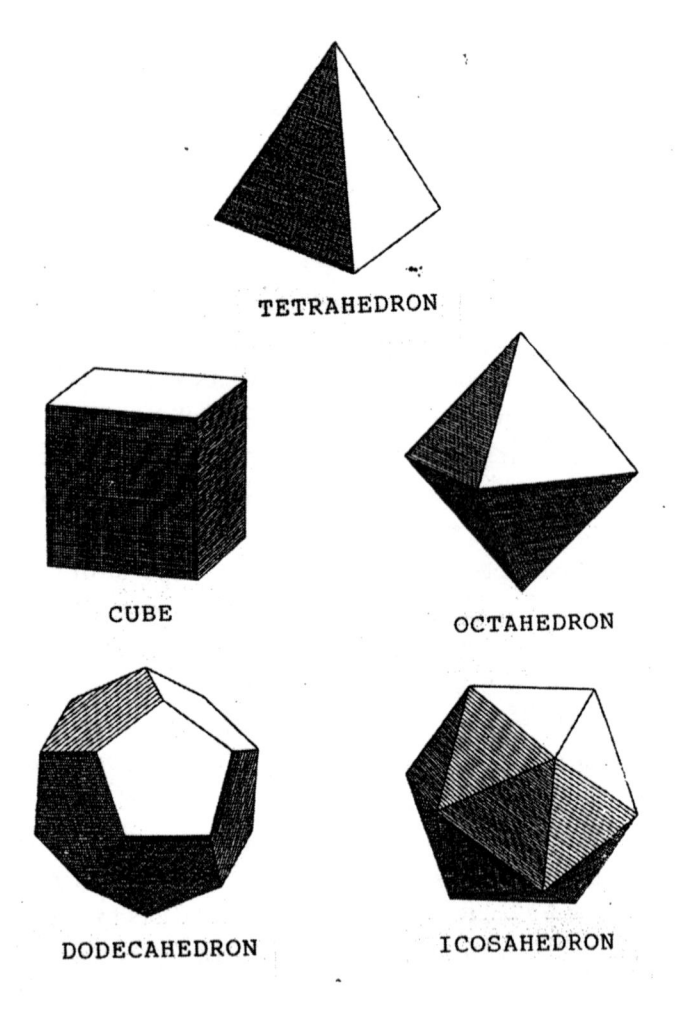

FIGURE 6 - THE PLATONIC SOLIDS

These shapes are easily seen in morphic resonance experiments and occur with different frequencies, a sample of which can be seen in Figure 7.

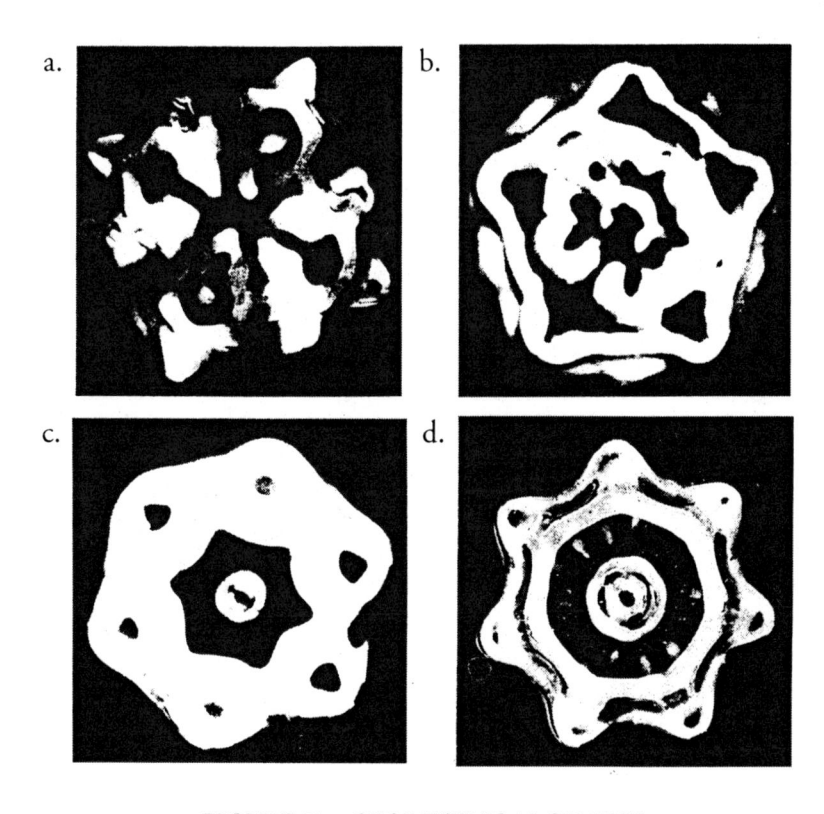

*FIGURE 7 - GEOMETRICAL SHAPES
CREATED BY MORPHIC RESONANCE*

The images (a), (b), (c) and (d) show four, five, six and seven-fold geometry respectively. They were created with liquid mercury at frequencies of between 130 and 240 cps. Liquid mercury, under high surface tension always seeks a spherical form. When vibrational frequencies were applied it did not cause disintegration of the drop of mercury but instead caused it to change form into these, and other shapes. The shapes at these frequencies were not static and were found to pulsate, with currents and vortices appearing.

Photographs from Cymatics: A Study of Wave Phenomena and Vibration, © 2001 MACROmedia Publishing, 219 Grant Road, Newmarket, NH USA. www. cymaticsource.com. . Used with Permission.

So geometrical shapes we can see, are inherent in the Om but the platonic solids have other qualities which makes them integral to the structure of matter. That is the golden ratio and this is described further in Chapter 20 The Golden Ratio and Fractals..

To further support the memory field hypothesis, take the example of homeopathy (which means treating like with like). In homeopathy a substance containing water and the remedy in question, whether plant or animal product, is continually diluted and succussed (shaken very fast). Eventually, the substance will have been so diluted as to consist only of water molecules, in physical, chemical terms, but has been imprinted with the memory of the original substance through the vibratory motion. The memory is in the form of frequency. So the water becomes the information carrier. The information being the frequency of the memory. It has already been discussed how everything has a frequency signature and it is the frequency which the water absorbs. So the water plus the memory field of the original remedy creates the homeopathic remedy and this is what creates the healing. It must be the case because, chemically, the substance contains only water. It also works on animals who couldn't be said to be using mind over matter. But, to assuage the sceptics, even if it *was* mind over matter, what does this tell us? That our consciousness is creating our reality, as already argued. The memory imprint was achieved through the increased vibration process and is consistent with the fact that faster speeds access higher (faster) dimensions. The memory field exists in a higher or faster dimension.

This increased vibration creates an increased frequency which creates access to higher (faster) dimensions of frequency. It is the same with increased spin of particles which also access the higher (faster) dimensions because the particles are taking in more light. A perfect example of this in everyday life is magnetic resonance imaging (MRI), a process which is in use in hospitals across the world to take a picture of the inside of the body. Walter Schempp, Professor of Mathematics at the University of Siegen in

Germany published a book in 1986 which proved mathematically that a hologram image could be produced from echoes of radio waves from radar. He called this wave holography. He realised that this is the same principle used in MRI. The MRI machine builds its picture by locating water molecules in the body. In order to see them it has to cause the protons of the molecules to spin faster than normal and does this by applying a magnetic field. So the increased spinning caused by magnetism creates increased resonance which creates the image. The pictures of the water molecules gives a picture of the soft tissues of the body. Now as the molecules slow down to their normal speed again after the image has been taken (it takes about 20 minutes) they radiate light. Schempp discovered that this light radiation contained wave information which the machine could use to build a hologram, or three dimensional image of the body. In other words all the information of the body was contained within the wave information or frequencies being radiated from the protons. He called this process quantum holography. This is a perfect example of the morphic fields existing in higher (faster) dimensions. All the information is contained there, which is a description of memory and the theory of Relativity or determinism which embraces all past, present and future, which of course would include memory.

So here we have a process which is in use everyday throughout the world which is based on resonance and which even accesses resonance of a higher dimension or frequency. The way it does this is by the increased spinning of the protons, which means they are taking in more light at a faster frequency. When the protons slow down their spin, they radiate the extra light which contains the higher resonance. This is a description of the phenomenon of baryons (the heavy protons discovered in particle accelerators, where protons are accelerated and accumulate excess light, described in Chapter 3 The Physics) and a description of the same process in biology called 'delayed luminescence' where a cell is subjected to high intensity light then after a short period of time it ejects the excess light to return to its normal state, or homeostasis, because particles and cells are in a delicate balance of spin and frequency.

A very efficient way of accessing faster dimensions is to cause particles to spin in vortexial motion centripetally (the opposite of centrifugally). In other words, spinning towards the centre. Viktor Schauberger was one of the first to realise this more by accident than design initially because he was trying to find the method of propulsion used by trout against a fast flowing current. He discovered that when water is propelled in a vortex motion, it creates enormous energy potential and this force is what propels the trout upstream or be able to stay motionless against a fast flowing current. However, in a vortex, speed is increased as the flow approaches the centre of the vortex. If speed is continually applied, as the flow reaches the centre it continues increasing in speed but since the centre of a vortex has a diameter of zero, this creates a translation, where instead of travelling in a spiralling motion with decreasing diameter, the molecules of water spin on their axis with increasing velocity. This increased spin allows the molecules and particles within it to increase in frequency because remember spin is linked to frequency (the photoelectric effect) so a faster frequency signature is attained. These faster frequencies exist in the higher harmonic dimensions or the zero point field. That is why some particles created in particle accelerators contain astronomical energy-units or gigaelectronvolts (like the baryons described earlier). That is because they have accessed the higher harmonic frequencies or zero point field through acceleration. Similarly, in homeopathy the water absorbs the frequency signature of the remedy which remains entrained within the water even if the chemical molecules of the remedy are removed from the water. It is a transference of frequency that has occurred.

Using this principle of vortex motion, any water can be purified by accessing a higher dimensional frequency. If water is contaminated by chemicals such as fluoride, chlorine, aluminium, they can be removed physically but the water would still contain their frequency signature. However, if it was spun through a vortex at high speed it would then become entrained with higher dimensional frequencies. The lower frequencies embed with the higher as the waves come into phase relationship forming coherent waves which access the faster morphic fields and are lifted or translated out of the physical. This is the principle of homeopathy; treating like with like. It is a higher level of the morphic field that is being accessed, or another way of saying it, a faster frequency of the same morphic resonance field.

Jacques Benveniste, the French immunologist, (1935-2004) came in for a clubbing from the scientific establishment when he dared to publish his studies, in the 1980s, on homeopathy which proved that water can hold memory. The scientific establishment, mainly Nature magazine, set out to disprove his experiments in a most unscientific manner. One of the 'fraud squad' team was a stage magician. The vilification against Benveniste tells us much about the factor of threat which the establishment feels when the status quo is being rocked and the situation is analogous to religious fundamentalism, where we see the backlash against the ordination of women priests by the catholic church, for example. For a time Benveniste's lab was closed down and his equipment confiscated but his experiments have been replicated by many other scientists since [2-5]. But what he found essentially is that atoms and molecules have unique frequencies. In experiments with dilution and succussion, where there was not one single molecule of a chemical remaining, it still exerted an effect. He discovered that the frequency of the chemical had been transferred to the water. This is homeopathy. However, in the early 1990's he went on to discover that he was able to effect molecular changes in cells by transmitting to them the *frequency signal only* of the chemical [6]. This led to an even more startling discovery. He digitally recorded the frequency, posted it by mail and e-mail to other parts of the world. When it was 'played' to a solution, the recorded frequency acted as if it were chemically present [7-9]. He called this 'digital biology'.

This principle of resonant frequency, however, had already been discovered approximately 90 years previously by Dr. Albert Abrams of San Francisco who taught pathology at Stanford University's medical school and was Director of Medical Studies. As well as being a top honours doctor he was also a master percussionist. His interest in using resonant sounds in his patients arose from witnessing the famous Italian tenor Enrico Caruso shatter a glass by singing a particular note. This note was identical to one which was produced by tapping the glass with his finger prior to singing it. He realised that sound could be a useful healing or diagnostic method. He experimented by percussing his patients and discovered that when any illness was present in any part of the body, the normally resonant sound changed to a dull tone. (This percussing is still in use by some doctors when they tap a patient's chest). This discovery led Dr. Abrams to develop what he called a 'reflexophone' based on a continuously variable electrical resistor which produced varying pitch of sounds detected from the wavelengths of diseased tissues. He later discovered that specific diseases could be detected and read from only a single drop of blood. During demonstrations of this to his fellow physicians he asked them for the scientific explanation of why they would prescribe quinine to treat malaria. They knew it was the cure but no-one knew *why* it was the cure. He placed a drop of blood from a malarial patient into the reflexophone whereupon it emitted a particular dull pitch of sound which could also be read on the dial. He then removed the blood and placed a few grains of sulphate of quinine into the machine whereupon it produced exactly the same pitch as the malarial blood. He then added the blood together with the quinine and the formerly dull pitch changed to a resonant pitch. He believed that the frequency of the quinine exactly cancelled that of the malaria and this led to his development of a machine which could emit frequencies to cancel out disease frequencies.

What he had discovered was the principle of homeopathy, which even today, still isn't entirely understood, but morphic resonance can explain why it works. What's happening is the combined frequencies embed and create coherence, enabling the higher frequencies existing in the morphic resonant field of the cinchona tree, which quinine is derived from, to be embedded with the lower frequencies of the disease, thereby translating the frequency out of the physical dimension into a faster, or higher, dimension. This is why the dull pitch changes to a resonant pitch. The frequency has embedded with the higher frequency and is in resonance with it. He called his machine an 'oscilloclast' (because it created oscillating frequencies) and by 1919 was training other physicians in its use. They used it to great effect on their patients although none of them really understood why it worked. When he published his work in the Physico-Clinical Journal in 1922, saying that he was able to diagnose patients from a single drop of blood using his machine, without the presence of the patient, this began what became vilification against him until the day he died and continued afterwards. The vilification came from the scientific establishment, what I have called up to now mainstream or orthodox science, mainly from the American Medical Association and the British Medical Journal. However, this view was not universal and glowing endorsement for Abrams came from Sir James Barr, former president of the British Medical Association who responded to the vilification calling Abrams 'the greatest genius in the medical profession'. Barr had successfully treated his own patients with Abram's methods. The full story of this and the upshot of Abrams work which has resulted in the science of radionics can be read in the bestselling book, already mentioned in Chapter 4, called *'The Secret Life of Plants'* by Peter Tompkins and Christopher Bird, published in 1973 by Harper & Row, U.S.A. Abrams treatment by the scientific establishment was echoed by Jacques Benvenistes' treatment by it, particularly 'Nature' magazine. Why this should be so requires some reflection but is most likely due to their own world view being threatened and is another example of how unscientific scientists can become when this happens. Abrams and Benveniste were following the scientific method and their work was replicated but their findings were obviously seen as a threat to the adherents of the prevailing world view, what I earlier called orthodoxy.

What their work is really showing us is that everything operates on frequency signatures. That frequency is even at the basis of chemical reactions and through the resonance of the frequencies we can see some evidence of the existence of morphic fields. Benveniste also discovered that water is the medium by which frequencies are transmitted throughout the body [10]. This was also the findings of Kunio Yasue, a physicist at the Notre Dame Seishin University of Okayama, Japan; that water somehow creates coherent photons [11].

Morphic resonance can also explain why it is that atoms stay where they are at all and don't just fly apart. Two Italian physicists, Guiliano Preparata and Emilio Del Guidice from the Milan Institute for Nuclear Physics have been trying to discover what it is that holds matter together, for example why atoms in any particular thing don't just float off into space. There is nothing in physics or biology to explain this. In biology again its back to the growth of form, why does something grow into the shape that it does and why does it stay that way? Morphic resonance again comes to the rescue here. Even for non-living organisms. When something is made, say a chair. It will then have its own morphic resonance field which is holding all the atoms in place, otherwise it couldn't possibly stay together. It is the sound which is holding it in place. And to be consistent with the theory already presented, that consciousness creates reality, it is the consciousness of the person making the chair which creates the morphic field (using thought, which produces the sound). They would be unaware of this because their consciousness is obeying the laws of physics of the third dimension which says that it is possible to make a chair. Nothing is made without someone first creating it with their consciousness, via thought.

But the curious thing these Italian physicists found when studying water is that, like light, it seems to form a coherent wave [12] and coherence enables information accumulation and non-locality. Considering that in each cell there are 10,000 molecules of water for each molecule of protein, it becomes understandable how instantaneous (non-local) communication is possible in the body. Stuart Hameroff, Anaesthetist and Director of Consciousness Studies at the University of Arizona, in doing studies into consciousness, including the NDE studies mentioned earlier, discovered that the microtubules within cells contain water which is ordered or coherent, [13 – 15] and that this ordered water produces coherent photons (particles of light), as Kunio Yasue discovered (remember coherence enables non-locality). Prior to this, from experiments he had already discovered that light is emitted from living tissue [16].

So it would seem that in the body, water is the method of transmission of frequencies, and that when water is coherent it produces coherent photons of light. It has already been stated that consciousness is light and coherence achieves non-locality, so that when light achieves coherence, this is the same as saying consciousness achieves non-locality. In other words, our consciousness is everywhere at once. Another word for this would be omnipresence, or what Buddhist's call enlightenment. Which means as humans, we have the potential to be enlightened or omnipresent. So, since it has been argued that everything is light, then consciousness cannot be confined to the brain, or even to the body. Many experiments have been conducted which confirm this finding. Experiments on distant healing and remote sensing/viewing confirm this and these are explained in later chapters, also as described in Chapter 11 Consciousness is the Key, PEAR labs at Princeton University, New Jersey have proved this in numerous experiments [17 – 22].

A scientist who has photographed and recorded coherent water or 'living water' as Viktor Schauberger would have called it, is Dr. Masaru Emoto of Japan. In his book *'Messages from Water'* he not only shows how water stores and transmits information but how it is affected by human consciousness. He has done this by taking photos of water crystals which, in healthy water, produces beautiful hexagonal forms, like snowflakes. In any water which has been subjected to stress or pollution for example, the crystals formed are distorted and incoherent. Even the intention which is behind certain words for example, seems to affect the crystals. In experiments with various opposing words, such as 'Love' and 'Hate', 'Thank You' and 'You Fool'. The words were typed onto paper and the paper taped to the water. In the former the crystal was perfectly shaped and in the latter the crystal was deformed, in fact it just looked like a splat. This was also true of the experiment where the words 'You Make Me Sick.' 'I Will Kill You', was taped on to the water. The resulting crystal was very ugly and distorted. The experiments were repeated using the English and Japanese equivalent of the same words and while the resulting crystals were different they were still either perfectly formed or deformed, consistent in each case. The experiments were also repeated using different technicians to account for the fact that an individual's consciousness may be affecting the results. This could still be the case, however, since it isn't clear from the book whether the person who taped the words onto the sample saw them. If they did, their association of the word could have transferred to the water. But this still shows us the power of consciousness in affecting the world around us and how careful we must be in choosing not only our words but our thoughts.

When you consider the violent language that is beamed out daily by television into people's living rooms, is it any wonder that the world is so sick? Dr. Emoto's work is giving us visual proof of the resonance created by the sounds we make from our speech and thoughts in this dimension. It is not only chemical pollution which makes us sick. We can also become sick from violent and unkind words and thoughts.

When you consider that coherent water enables light coherence, then it becomes necessary to have coherent water, or healthy water in order to have not only general good health, physically, emotionally and mentally but also coherent consciousness which can eventually become non-local leading to the state of awareness of the Whole, also known as Self-realisation, Omnipresence, Enlightenment. With the pollution of waterways around the world it is becoming harder and harder to have healthy water. However, all is not lost. There are ways of returning water to optimum health. There are various technologies available to achieve this. The Centre for Implosion Research in Plymouth, England, [23] have built vortex spirals to cleanse water and increase its energy field. The spirals are fabricated to the golden ratio and filled with imploded water. That is water which has been caused to spiral in a vortex fashion which, as explained, enables it to attain a higher (faster) dimension. The water is then entrained with this frequency and experiments have shown that when placed next to disordered water, the disordered water becomes entrained with the ordered or coherent water (the same principle described earlier about homeopathy, where the disordered frequency is embedding with the ordered frequency). The work is based on Viktor Schauberger's findings and the vortex spirals can be placed on the incoming water pipe of a house to cleanse and transfer higher frequency energy into the incoming water. Johann Grander of Austria has also built technologies based on the vortex spiral principle to increase the frequency of water, although as of the time of writing I don't know where it can be purchased from. Emoto's work also shows us, however, that we can make our water ordered and coherent by speaking loving words or thinking good thoughts towards it. This is possibly the reason for the universal practise of saying grace or blessing our food before a meal. So our consciousness is affecting everything and this is what quantum physics has told us and also what spiritual philosophies have told us already. That we can change reality with our consciousness because our consciousness creates reality.

    Masaru Emoto's work also shows us that thoughts and words of love and compassion create coherent patterns. Because love creates ordered and coherent water crystals. Water is the information carrier for frequency in the body. When the frequency from the water crystals is coherent, this creates coherence of lightwaves. As greater coherence is reached this enables greater contact with the higher (faster) morphic fields of consciousness. Through love and compassion we grow in evolutionary terms. This has been lost to a world which is obsessed with technological fixes, which values the intellect above love and compassion. But to damage and destroy the means necessary for survival as is happening in the world today, is neither compassionate nor intelligent. Intelligence is already built into the morphic resonance fields. We have to find the way to reach them, or embed within them and Emoto's work is an example of present-day scientific proof that Love is the way to achieve that coherence. We can also look to other evidence of this same fact. Every spiritual tradition teaches that love and compassion are the number one law: spiritual healing is practised the world over and is an ancient practice and it works by transmitting love, seen as light, which enables a person to come into a state of coherence and in our own lives we <u>know</u> that love and compassion heal us.

When the blue tits made their discovery they lifted the species into a higher (faster) level of their morphic resonance field which was already in existence, as the evolutionary stimulus will be built into the OM. Each morphic field is like an onion, with many layers, each layer having faster frequency but the template is self-similar throughout. As described earlier, at increasing harmonic speeds of light there is a greater and greater consciousness where all information is known. This is why dowsing works and why intuition is always correct. Any new discovery or invention, in any of the kingdoms, already exists in the template of the OM in one or other of the dimensions. This would account for the fact that a new invention is often discovered by several people at the same time, as any patent office will tell you and often the idea comes to them in a dream. As with Carl Jung's book described above, Thomas Edison (1847-1931) the inventor of the electric lightbulb, also dreamed of his invention. The information is sitting there waiting to be discovered. This must be the case as relativity theory states that past, present and future co-exist, which means that every new discovery and invention is already in existence somewhere. When the timing is right, it is 'discovered' (this is quantum physics, the observer creating) and the consciousness is lifted, to a higher or faster speed. The body and consciousness will then be resonating to that higher or faster morphic field. This explains evolution and also the 'blue tit example'. It is also another example of relativity and quantum theories being part of one law.

In ancient wisdom teachings it says that the mineral, plant and animal kingdoms have group consciousness because they are part of a group soul, e.g. in the animal kingdom there is a cat soul and all cats are part of that one consciousness pertaining to cats, all birds are part of one bird soul. This would explain why birds flying in a group can turn and move all at the exact same time as if they are one mind. Fish do the same. These teachings also say that it is only humans who have individuated souls because it is through the human kingdom that matter is spiritualised (although we and everything else are ultimately from and interconnected with the same source always. There is no reality in separation, even with individuated souls. Individuated souls are still part of the one collective consciousness of which everything is a part). But human beings, as well as being individuated souls, *believe* in separation and that's probably why we're so clumsy and bump into each other.

The image of the vortex spiral is a useful way of visualising the process of evolving consciousness. The evolutionary impulse can be likened to a turn of a spiral, the outward cycle would be a gradual forging of autonomy, from group soul through mineral, plant and animal into the individual human soul. At some point in the human consciousness in one or other life a pull is felt, towards something greater, something beyond ourselves, a striving for connection to a deeper meaning. This then begins the return home to Source and the human consciously treads the spiritual path (of which there are many). They are literally being called home and beginning the second half of the spiral taking them back to Source, but with experience, or in other words, evolutionary gains, or evolvement of consciousness. This can be visualised using the conical vortex model. The 'pull' that is felt is similar to the pull that people feel in near death experiences when they feel drawn or pulled towards the tunnel of light. Remember that a vortex naturally draws all towards the centre.

Through a succession of conical vortexes over eons of time, perhaps there is a new big bang to begin a new cycle or period. In ancient Hindu philosophy they talk of Yugas, which are cycles of time, each lasting for hundreds of thousands of years and they are repeated over and over into eternity. This is what was referred to earlier as the in breath and the out breath of Source. Perhaps this is what the expansion of the universe is, the in breath, or even the out breath depending on how you look at it. Followed by implosion, followed by expansion etc. etc. Considering that OM is sound, this would imply an out breath, so perhaps what we have now is the out breath of Source. When a string of conical vortexes are placed end to end it begins to look like a never ending spiral and could be spirals within spirals, analogous to strands of DNA.

Many scientists believe that ultimately the answer to the mystery of form will be explained by DNA. My view is that the answer may well lie in the DNA but not in the way they believe i.e. through the physicality of chemicals and genes.

DNA is a chemical substrate, but like all matter, it is made up of sub-atomic particles, each a tiny vortex, as already explained. The DNA itself is a double spiral. Again we are back to spirals and 'as above, so below'. It is also important to think of the analogy here; the DNA (spiral/light wave) determines if the cell (vortex particle) will change or stay the same. As explained, all form is in a state of potentiality. So the DNA spiral can keep the existing form of cell or change the form of that cell to fit the overall geometric pattern but it is the sound from the morphic resonance field that determines the form. The 'potential' form already exists in the sound, the OM, at a very high frequency. Consciousness via thought, or the right stimulus to use Schauberger's phraseology, (which is quantum physics), brings the particles into physical existence and the form takes shape to fit the pattern created by the sound. As this process continues endlessly and always has done even without human intervention and hence human will, it therefore means that a higher will is driving evolution. This is probably the strongest argument for the existence of a Creator Intelligence. But because we have creative will, humans have discovered ways to manipulate the form using genes, without the basic understanding of the forces at work, the Plan, the Harmony or Divine Will. This is leading to a very dangerous situation on Earth today. However, if carried out with altruism and awareness it could bring tremendous benefits to humanity and may eventually lead to replacement organs being created from a person's own cells which would prevent the problem of organ rejection.

Now I want to give a perfect example here of 'potentiality' to use Viktor Schauberger's very apt word, which shows, apparently, the wave/particle duality of light, but actually shows the evidence of morphic resonance at work as well as proving that light has consciousness, or perhaps more correctly to say, light is consciousness. It is the double slit experiment taught to all undergraduate physics students.

CHAPTER 14

THE LAW OF RESONANCE

> *"Professor David Aldridge at University of Witten Herdecke in Germany suggests that disease is a type of musical dyslexia and the inability to "keep our rhythm" while living with and being bombarded by the varied rhythms of the world around us."*
>
> Paul Pearsall, Medical Doctor, Speaker and Author (extract from 'The Hearts Code')

The double slit experiment used to be carried out to show that light sometimes behaves as a wave and sometimes as a particle though most often nowadays it is used to show the weirdness of the quantum world. It is set up to enable a beam of light to pass through a plate which has two equal-sized slits on it. When the beam of light is shone through the slits an interference wave pattern is displayed on the other side which is exactly what is expected to happen. The interference waves create bands of light on a back plate. This same interference wave pattern occurs when water flows through a similar double slit. Two waves colliding will produce an interference wave.

The interesting part of the experiment happens when instead of a full beam of light being directed at the plate, a series of single photons (particles of light) are directed towards the slits. Logically, it should be the case that a single line should appear behind each slit because there are no other photons to cause the interference wave, however, what happens is that eventually, when enough photons have been sent individually, they form the interference wave bands of light as with the full light beam. This happens even if there are long gaps of time between each photon being sent.

This is a complete puzzle to science as the laws of physics say that only two lines should appear, just as if single particles of sand were dropped through the slits onto a plate below, two lines of sand would eventually appear, there would not be interference wave lines. Then an even stranger fact was discovered. Again, when sending single photons, when a detector was set up next to the slits to detect which slit each particle went through, the interference lines disappeared and only two lines appeared. So the act of observing (even if it is a machine) stops the interference waves of single photons.

This experiment was first conducted in 1909 by Thomas Young and the explanation at the time was that it was because light is both a wave and a particle but this is not a good explanation. It is simply a statement. In the 1920's a theory was born to explain it called the 'Copenhagen Interpretation' because the scientists involved were based in Copenhagen. The theory stated that particles, or photons, can be in two places at the same time, that they create a double of themselves and pass through both slits at the same time as long as it is not observed. If it is observed, then it chooses one position over another and the detector detects it passing through one slit only. The theory also states that this dual existence only applied to the microscopic world and not the larger world of which we are most familiar. This in itself should have been very suspect, since the larger world is made of the microscopic world, so the same laws must apply.

But in 1957 another theory came along to attempt to explain it, known as the Many Worlds theory, proposed by Hugh Everett of Princeton University. This states that there can be multiple realities and when a particle has to make a choice then another reality comes into existence. The two realities then interfere with each other even across great time gaps. It states that a particle from a parallel reality can come back and interfere with this reality. So, a particle on one reality goes through one slit and a particle in another reality goes through the other slit even over time gaps, and they somehow interfere with each other (causing the interference wave). The theory goes on to say that this only happens in a limited way with no consequences for the everyday world. This again is the flaw in this theory but as well as this there are simply too many caveats.

Dr. David Deutz of Oxford University also believes in the Many Worlds theory or 'parallel universes' as he calls it and he believes that this is the only possible explanation for the double slit experiment. His theory differs slightly from the Many Worlds in that he believes all the realities co-exist concurrently. In other words, they don't just come into existence when a particle has to make a choice.

Another theory proposed by Louis de Broglie and David Bohm, known as the de Broglie-Bohm explanation, is that the particle or atom always behaves like a particle and only goes through one of the slits but what they called 'the quantum potential's influence' is spread over both slits and guides the atom along one of the wave pattern paths. They never could explain what exactly the 'quantum potential influence' was, however. But I believe they were on the right track as this 'quantum potential influence' is what I am calling the morphic resonance field.

I don't believe any of the above theories *adequately* describes the double-slit experiment as they are only believed to work in a limited way to explain the quantum world and have no consequences for the everyday world; and there is no explanation of the actual physics involved. In fact this is true of quantum theory per se. It is not understood how the particles become particles. It is also not understood why the atom has the structure that it does. In theory, the negatively charged electrons should continually be spiralling in towards the positively charged protons in the nucleus. The weak nuclear force is used to describe this mysterious force which keeps them apart but science does not know what this force is. Similarly, it is not known why the protons and neutrons in the nucleus are so tightly bound together, this they call the strong nuclear force.

Morphic resonance can explain all of the above and I would reiterate the Occam's razor law quoted in the introduction. Morphic resonance is the simplest explanation and describes how particles form, how they create the structure of the atom and, in turn, create larger forms and how the double-slit experiment creates an interference wave no matter how many particles are sent through each slit or at what proportion or time frame. The particles are taking their place in an existing pattern, that is the morphic resonance field of the form (in this case the plate).

In the first part of the experiment a light beam is directed towards the double slit. A set of interference lines shows up on the back plate as expected because waves or periodicity, as already explained, is inherent in all of nature. The light is forming the shape suggested by the double slit. It's simply conforming to the physical laws of the double slit. What is not so obvious is that there is a greater law at work which is holding the physical laws in place, a greater pattern which all physical life is resonating to. It sits behind the physical laws and the double slit experiment demonstrates this beautifully. This doesn't become obvious until the second part of the experiment where single photons are directed towards the slits, one at a time. Again the interference pattern shows up because the morphic pattern of the double slit plate is resonating. The fact that you have a double slit plate means that there is a morphic pattern resonating for it. So the pattern remains because it is an existing pattern and will remain fixed unless the resonance of the actual plate changes. The resonance wouldn't be changed by the single photons because, remember, the photons or light waves take their place in the pattern created by the resonance (of the plate with the double slits) and not the other way round. **This law I am calling the Law of Resonance.** The pattern is fixed because it obeys this law. The double slit experiment shows that the single photons are still obeying this law. They have no choice but to keep this natural order because it is a law of resonance. When this changes is when the particles are observed (either by human or machine). They then obey the law of the physical dimension which says that single particles do not cause interference. These laws are self consistent with the behaviour of particles and waves.

If anything, the double slit experiment shows us that there are indeed other unseen realities having an effect on this reality. The many worlds or parallel universe ideas are not too different from morphic resonance in the sense that they are saying that many unseen realities co-exist. Indeed you could say that the higher morphic fields are parallel universes. Morphic Resonance is a more complete theory, however. It is saying that the different realities are fractals which are based on different (harmonic) speeds of light. This would explain why the patterns are consistent (such as the double slit plate) and why they are unseen (dark matter). It also models beautifully the inherent pattern of periodicity of waves in all of nature. **It also proves the central theme of this book: that light is consciousness.** When a particle or photon (both light) is being observed by this dimension it obeys the laws of this dimension and reverts from creating interference waves to creating two single lines, which is what happens when single particles are sent through two slits. It is a law of the physical dimension. This happens no matter how the observation is carried out, whether by people or machines. The photons and particles cannot be tricked. **They know when they are being observed and they make a decision. This proves that they must have consciousness because consciousness is the prime requirement for decision-making.** Since photons (particles of light) and other particles behave the same way they must both have consciousness which proves that particles must be made of light, as already argued .

To summarise the truths of this remarkable experiment we can say that:-
a) morphic resonance is the template for all matter
b) photons and all particles are made from light and that light is consciousness
c) consciousness obeys the rules of each dimension

Another experiment known as the 'phantom leaf effect' is also proof of the morphic resonance fields, using Kirlian photography. This was discovered in the 1940s, invented by Russian scientists, Semyon and Valentina Kirlian. It was suppressed for many years as it didn't fit the conventional scientific understanding but as this is gradually being expanded, it is now being used by more open-minded scientists. It is an excellent tool to show proof of the morphic resonance fields of which physical matter is a precipitation. The famous 'phantom leaf experiment' consisted of cutting off the tip of a leaf which, when photographed using Kirlian photography, showed the entire image of the leaf intact, despite the missing part of the leaf having been destroyed. This was not due to remaining moisture or some such substance remaining on the plate as originally thought by some sceptical scientists. It's origin was described as an unknown energy field

As with the double slit experiment, the morphic field remains because it is following the law of resonance of the physical dimension. The morphic field remains despite what happens to individual particles. Indeed the particles are compelled to take their place in the pattern. This means that one day it will be possible for humans to do what salamanders do; grow new limbs or body parts, as the template (morphic field) is intact. This would also explain why amputees still 'feel' the presence of their missing limb.

Proof of the morphic field was given by the Professor of Anatomy at Yale, Harold Saxton Burr (1889-1973), in experiments he carried out in the 1940s. He discovered an electrical field around living organisms which was the same shape as the organism but the most interesting finding was that this field existed in its complete form even when the young organism was not yet fully formed. For example, the electrical field around salamanders was always in the same shape as the adult salamander. This was true even in young salamanders and even in their unfertilised eggs [1]. This was also true for plant seedlings whose field was the shape of the adult plant. These are perfect examples of morphic resonance fields. They are already in existence and the organism grows into them. The sub-atomic particles have no choice but to take their place in the resonating template. This is the explanation of the growth of form, which cannot be explained by DNA alone.

Another example of morphic resonance is quantum computers. These computers are able to perform almost instantaneous calculations, in fewer steps than logic would allow. Scientists working with these are talking of parallel universes as a way of explanation. The idea being that the answers already exist in a parallel universe and this is somehow being accessed and I believe this is indeed what is happening. The atoms in the computer are accessing the higher, or faster, morphic resonance fields, by different methods because there are several technological models being experimented with. In these fields all information is known, thereby by-passing the logical steps of computation. Dowsing, or divination, as described earlier, is also accessing these same dimensions, essentially via speed, though there are different methods by which to achieve a faster speed of consciousness.

Resonance is a fact of nature. In the everyday world we see it at work and we use it to our advantage. For example lasers, an acronym for Light Amplification by Stimulated Emission of Radiation. This means that, normally, particles of light are released from atoms randomly but lasers require a steady supply of atoms to release their particle of light at precisely the right moment to create a coherent wave of light. In other words all the particles of light are released at the same time producing a coherent (not scattered) beam of light. This is achieved by submitting an atom to a particular frequency. The particle of light stimulated by this frequency will cause all other particles within its vicinity to exhibit the same frequency. In other words, they resonate together. This is the same phenomena that occurs when pendulum clocks are placed near each other. The most powerful pendulum swing will gradually bring all other pendulums swinging to the same frequency. This is the same as saying the most dominant resonance creates coherence.

This also means that sound will be one of the greatest healing methods in the future. Disease is nothing more than disharmony or dissonance. Being out of tune. When a person can be brought back into the correct resonance, then their disease, or dissonance, will be cured. This is why music can be so healing and why it can lift us and move us to tears. When music has this effect on us it is because we are in resonance with it. This is the principle of healing. When healers tune into higher dimensional light, they are really practising morphic resonance. The light, remember, contains within it the sound /resonance.

CHAPTER 15

PROOF OF SPEEDS FASTER THAN LIGHT

> *"...what we call scientific knowledge today is a body of statements of varying degrees of certainty. Some of them are most unsure; some of them are nearly sure; but none is absolutely certain."*
> Richard Feynman, Physicist (1918 - 1988)

This dimension we live in is usually called the third dimension, which describes, length, breadth, height, but it is actually four dimensional if you include time. This is what is commonly referred to as space-time in physics. So beyond the speed of light you enter the fourth dimension, or fifth, depending on how you look at it. Time still exists in this dimension but is slower and outwith the range of our five senses to measure. Remember that relativity showed us that travelling faster produces slower time. Time is slower but the vortices/particles/atoms of that dimension are spinning faster. This process continues into each harmonic multiple of speed which creates ever-ascending dimensions. Each dimension having a faster spin of vortices/particles/atoms but slower and slower time, until the ultimate dimension is reached where there is no time because it has stopped altogether. This is the ultimate description of relativity/determinism.

This is also a description of the spiritual realms. We cannot see them because they are moving too fast. The key to being able to see these realms is if we can 'raise our consciousness' in other words 'tune in to a faster speed'. This is what is happening with the quantum computers mentioned in the last chapter, or with dowsing. This is also what is happening with people who are clairvoyant or clairaudient and it can happen to anyone on a spiritual path because they are using their will (via various practices - prayer, mantra, meditation, in other words, inner work) to raise their consciousness. It enables them to 'tune in' to higher speeds. This should be no more difficult to understand than tuning a radio set to a different frequency. The frequencies are all around us, passing through us all the time, but we are unaware of them unless we 'tune in'. We cannot hear the frequency unless we tune in. So the higher levels of consciousness are functioning through ever increasing speeds, invisible and inaudible to us, unless we know how to 'tune in'. That's why we talk about 'raising our consciousness' or being 'lifted up' - not to a place but to a faster speed. A faster speed of spin of particles and hence a faster frequency.

I deliberately talk of frequency for radios and speed for consciousness because frequency refers to the number of waves per second which is how radio frequencies work but when I speak of speed I am referring to the rate of spin of the vortex/particle on its axis. But the vortex/particle, while spinning, also vibrates to a frequency so it has both rotation and vibration, or another way to say it is, it is spinning around and jumping up and down at the same time. The rate of frequency is necessarily linked to the rate of spin which is the same as saying that the vibration is linked to the rotation. This was discovered by Einstein in an experiment known as the photo-electric effect. He concluded that the energy of light particles fixes the rate of vibration. In effect, he proved that the rate of spin is linked to the rate of frequency. He was dealing of course with the speed of light of this dimension but the same principle applies to other dimensions, other harmonic multiples of the speed of light, due to the fractal principle. If the particle was to increase frequency without increasing spin, it would cook. This is the principle of cooking; applied heat is an increase of frequency. The increased frequency of waves increases temperature and physical life is highly sensitive to temperature changes.

Below is physical evidence of the effect of increased frequency (temperature) on the morphic resonance field. On the left, the sub-atomic particles (of the sand) are spinning relative to the appropriate frequency. On the right, the temperature (increased frequency) causes the speed and frequency to be temporarily out of sync, creating a distortion of the pattern.

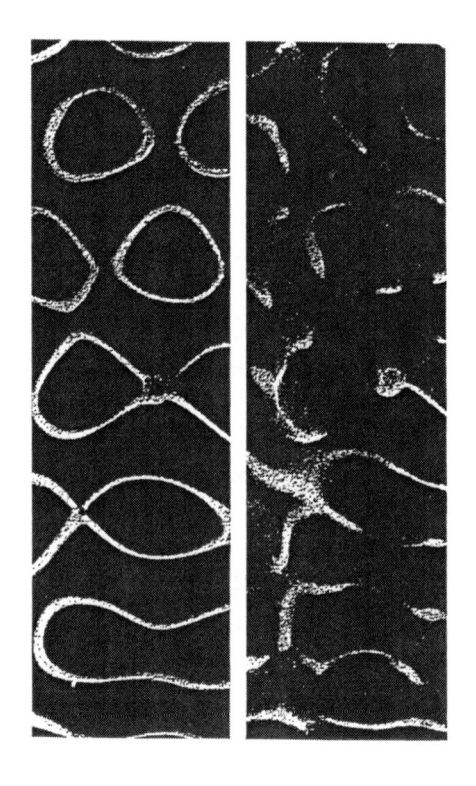

*FIGURE 8 - MORPHIC RESONANCE
FIELD DISTORTED BY HEAT*

Figure on left: Plate 24.5 x 32.5cm, thickness 0.5mm, 1580 Hz before heat was applied.

Figure on right: Same plate after a corner of the plate was touched by a flame for a few seconds. Figure on left reappears once plate has cooled.

Photographs from Cymatics: A Study of Wave Phenomena and Vibration, © 2001 MACROmedia Publishing, 219 Grant Road, Newmarket, NH USA. www.cymaticsource.com. . Used with Permission

So when particles spin faster, time alters, slows down, as demonstrated by the Special Theory of Relativity. When something moves very fast in the physical dimension it seems to disappear (e.g. helicopter blades spinning). It is the same with the higher / faster dimensions. The only reason we can't see them is because they are vibrating so fast in frequency, but to them, at their rate of spin of particles (which has slowed time) time seems normal, frequency seems normal and they are not cooking! Once again, it is their faster rate of spin which has slowed time which means they can exist in a faster frequency without overheating. **The rate of spin of our particles determines which frequency we can exist in because it's the rate of spin that determines time, the rate of spin being affected by the speed of light waves entering the vortex.** So ultimately it comes back to the speed of light being the determining factor.

This also explains why Earth time is seemingly different than time in the spirit world. The different dimensional octaves allows for time differentials. This also explains time travel which Relativity theory predicts. Travelling faster produces slower time. If a person travelled away from the Earth at the speed of light, their time would slow down. If they were gone for a year according to their clocks, to them it would seem that a year had passed, to those back on Earth, it would seem that ten years had passed, or even 100. The time differentials could be worked out as they would be linked to multiples of the speed of light.

Relativity means that space and time are relative (not fixed) and time slows down when approaching the speed of light. As already noted, when speeds greater than the speed of light are reached time would logically have to slow down to the point of stopping altogether, so past, present and future would co-exist. Physically, we may or may not be able to travel faster than light but Life is not just physical as I have already argued. Consciousness is not physical and speeds faster than the speed of light are possible. Professor Gunter Nimtz of the University of

Cologne has transmitted music (a Mozart piece) at a speed of 4.7 times light speed at a distance of 14 cm.

Much like Marcus Chown mentioned earlier, who said it was possible to send a photon faster than light but it would not contain meaningful information, the physicist, Professor Raymond Chiao of the University of California has, through experiments, concluded that it is possible to send the occasional, random photon at faster than light speeds. One experiment achieved a velocity of 1.7 times the speed of light. Where Professor Chiao, Marcus Chown and others part company with Professor Nimtz is when he says he can send information in this way. Professor Nimtz did indeed send the Mozart piece as described, but Professor Chiao and other scientists are arguing about what constitutes a signal and whether Mozart is information!!

Let's look at some evidence:-.

1. Professor Chiao and his colleagues admit, and have proved, that faster than light speeds exist.
2. If one photon can travel faster than light then so can *all* photons.
3. Music is wave frequencies. It is most definitely information.
4. It has been argued that particles at very high frequency contain an accumulation of information.
5. The case could be very simply decided if the experiment was replicated by the critics but so far they have failed to do so. Perhaps because it violates their beliefs about what is possible. Unfortunately, this unscientific behaviour doesn't help the advancement of science.
6. Professor Chiao and other scientists' insistence that faster than light signalling (the sending of information) is not possible, is based on the belief of the probabilistic nature of the quantum particles. But 'probability' of quantum particles is a theory only and I am arguing that probability

or chance is not the accurate description of quantum particles, rather they operate on 'potentiality' and obey the morphic resonance templates. So faster than light signalling is allowed when probability is replaced by potentiality.

Relativity/determinism has also already pointed to speeds faster than light because if past, present and future co-exist and there is no time as determinism says then time must have stopped which means that faster than light speed must have occurred. This is a logical conclusion. It has already been proved that faster travelling produces slower time so eventually faster speeds must stop time. There is no point arguing over this if we accept determinism which means accepting that past, present and future co-exist, then we must accept that speeds faster than light must have occurred.

Further proof that faster than light speeds are possible and that light speed is not constant is from the study of frequency. All energy vibrates at different frequencies. The frequency is the number of wavelengths per second. So within that second of time, there could be 10 or 10 million wavelengths. In order to reach that one second of time slot, the 10 wavelengths would travel relatively slow in comparison to the 10 million wavelengths which would have to travel very fast. Think about it! So the speed of light cannot be constant. Also if we remember the evidence that lightwaves are actually spirals, this means that frequency will be dependent on the speed of the angular acceleration and radius of the spiral. So within frequency ranges we have a non-constant speed of light.

Also, as described in the chapter on The Physics, the electromagnetic spectrum is a bandwidth of frequencies which we can measure with our instruments but has no set limits either end. It is open-ended and has been described earlier as a subset of a larger spectrum, of unknown limits. This means unlimited frequencies with unlimited speeds of light. As spin is linked to frequency, this means that particles can exist in frequencies outwith the known bandwidth. This would explain quantum events, the zero point field, the dark matter and spiritual phenomena such as dematerialisation and re-materialisation. However, as described at the end of Chapter 2 The Misunderstandings, the experiments on teleportation have now brought de- and re-materialisation into the scientific domain. Dematerialisation occurs when technology or consciousness is used to increase the speed of spin of particles which then promptly disappear from the third dimension because they are beyond the speed restrictions of the physical dimension. They will be spinning faster and vibrating to a faster frequency which makes them disappear (like the helicopter blades) although this analogy is still describing a third dimensional object which still exists in the third dimension but I'm using it to show the principle of speed creating invisibility. When you increase in speed beyond the spectrum of the (spin and frequency) of the third dimension you physically do not exist in the third dimension but you do exist in faster dimensions. If you are having trouble with this concept, remember what was explained earlier about the ethereal nature of matter, that each atom of matter is made of mostly space, that matter and space are all light and that each sub-atomic particle is simply spinning light (or spinning energy if you prefer the more conventional explanation). So all the dimensions are co-existing and it is simply a case of 'tuning in', rather like tuning a radio to the correct frequency. When the speed, and resultant frequency, is lowered again, the person or object once again appears in the third dimension. This explains the appearance and disappearance of many phenomena which people all over the world experience, such as angels, spiritual figures, UFOs and just plain objects. Many millions of people all over the world have experienced such phenomena that it cannot be dismissed, indeed many by multiple witnesses so it cannot be explained away as psychological disturbances or hallucinations.

So this dialectic theory states that beyond the physical speed of light there are harmonic multiples of the speed of physical light. This is the only assumption in the theory as stated in the introduction. I call it an assumption but there is already proof of faster than light speeds. I am, however, saying that they are 'harmonic' speeds. The argument being that as all life is based on sound and sound follows harmonic octaves, then it is reasonable to assume that light also follows harmonics because remember harmony means 'progressions but all simultaneously sounded and the combination of simultaneously sounded musical notes to produce chords and chord progressions forming a pleasing and consistent whole, which is a not only a description of the OM but provides a consistent and simple theory. Also, as already explained, sound and light are the same thing relative to speed - light is fast sound, sound is slow light.

CHAPTER 16

ELECTROMAGNETISM

"Miracles do not happen in contradiction to nature, but only in contradiction to that which is known to us in nature"
St. Augustine (354 - 430 A.D.)

Electricity and magnetism are implicit in all life, from the atom right up to the planet itself and all life on it. Every chemical reaction in the body creates an electrical force. Michael Faraday, the 19th century scientist who discovered the principles of electromagnetism, said of his own discovery "the atoms of matter are in some ways endowed or associated with electrical powers, to which they owe their most striking qualities, amongst them their mutual chemical affinity." Each of the electrons, positrons and protons in the atom carry an electrical charge, the electrons are negative, the protons and positrons are positive. Positrons (anti-electrons) come from cosmic rays, and protons (which create the little neutrino baby with the electrons) are part of the nucleus We are continually exchanging electrons with everything around us is due to the electrical attraction of positrons and protons to electrons, the attraction of positive and negative charges. An obvious example of this is when you feel an electric shock from static electricity, it is an exchange of electrons that has taken place. This means everything is dynamic, nothing is ever truly at rest When electrons meet positrons, this is matter and anti-matter, they annihilate each other, in a flash of light. It's really just to do with opposite electrical charges cancelling each other out. The electrons which don't get annihilated by the positrons are attracted to the only other positive charge around which is the proton. So some electrons will be annihilated, some will mate with the proton creating life. Rather analogous to the sperm and egg situation. Throughout the whole dynamic exchange an energetic balance is always maintained to enable atoms to exist and to create new life.

Science has been able to measure the effects of electricity but no-one really knows what it is. It is one of those mysterious facts of life, like consciousness. It is obvious they exist, yet what are they? It is my understanding that electricity is life energy itself, a projection or an influx of the light waves of consciousness, which creates a change in the morphic resonance field. If we remember right at the beginning how the light is the breath of the Source, and the sound of the light is the morphic resonance fields, so the light and the sound are essentially the same thing, sound is slow light. We have already said that breath is synonymous with spirit so we can say that whatever light and sound are they are synonymous with spirit. This means that in essence, everything is spirit, or everything is light and sound, it's the same thing.

What does it mean to say that light and sound are the same thing? Sound, like light, are both part of the electromagnetic spectrum. And like the rest of the electromagnetic spectrum, have both frequency and wavelength, so the same unit (Hertz) can be used to measure all of it. Hertz is just the number of cyles per second or waves per second, frequency in other words. So light and sound are the same thing relative to speed. Remember light is fast sound, sound is slow light. As for saying that they are the same thing, it's perhaps more true to say they are a reciprocal pair. In grammar, reciprocal means *'expressing mutual action or relationship*' and in mathematics means '*a quality or function related to another so that their product is **unity**'* ... Again, It's all **Light**. The truth of sound and light being the same thing has always been known as part of ancient spiritual teachings but if modern day proof is required this can be demonstrated on a computer. The musical notes we are familiar with in an octave – F,G,A,B,C,D,E,F – if played 40 octaves higher, will produce the colours of the visible spectrum. So F produces red, G produces orange, A produces yellow, B produces green, C produces blue, D produces indigo, E produces violet and F (upper) produces magenta, also known to us as ultra-violet. Another way to describe this reciprocity is, 'sound is slow frequency light, and light is fast frequency sound'. This is very amply demonstrated in a book called 'Civilization One' by Christopher Knight and Alan Butler which they call 'the spectrum from music to light'. It's important to realise that this example is only one tiny sliver of the electromagnetic spectrum, most of which is invisible and inaudible to us. So we are both blind and deaf to most of reality! This is worth pondering on. It exists, despite the fact that we are blind and deaf to it – unless we can tune in. We have technology to tune in to extreme ends of the electromagnetic spectrum, with x-ray machines at one end and radio sets at the other, but the spectrum doesn't end there. That's just the limit of our present technology and ability to measure. So, it's all a case of tuning in. The universe could be likened to a gigantic transmitter and we, in our normal human state, have the equipment (the 5 senses) to tune into a particular station. However, it has been argued and will be explained in more detail later, that it is possible to tune into other stations, either of our familiar electromagnetic spectrum or harmonic octaves of it.

The morphic resonance fields are the sound of the light, so we could say they are spirit, or exist in spiritual dimensions, remembering that spirit is just unseen light. As there are harmonic dimensions of light there would necessarily have to be harmonic dimensions of the morphic resonance fields and although these tend to be unseen, some of them, perhaps those relative to the third dimension, can occasionally be detected by equipment. We wouldn't be able to see them with our eyes (unless we know how to tune in) but x-ray equipment has the potential to pick them up. I have heard suggestions that this has actually occurred, where x-rays of amputees revealed an image of the limb still intact. This happens with Kirlian photography. If you recall earlier the description of the phantom leaf effect and also the findings of Harold Saxton Burr and his electric field templates around plant seeds and eggs which the organisms 'grow into'. This suggests that the morphic resonance fields are a form of electricity and this would be consistent with my earlier explanation that electricity is life force or lightwaves of consciousness. Electricity in the form of the morphic resonance fields would be slower light; sound in other words, as we've already seen that light and sound are the same thing relative to speed and are synonymous with spirit – so we have different names for the same thing Spirit – matter, are descriptive in physics terminology as lightwaves – vortices and are essentially the same thing, all light, but we see the duality of spirit and matter because of the level of our consciousness. This duality is seen from the smallest scale up to the largest. The spiralling lightwaves creating vortices in the sub-atomic world, up to human reproduction, where the sperm (from the male) enters the ovum (from the female). The sperm even behaves in a wave-like fashion with its wavy tail, just like the spiralling light wave. The ovum is circular and receptive like the vortex particle. Right up to spiralling galaxies creating clusters of stars.

Now the modern-day evidence for electricity being a light wave of spirit is from the work of cloning. In cloning, the egg only begins to grow when subjected to a burst of electricity. It is the electricity which begins the life process, analogous with the sperm, but no sperm is involved.

This is in keeping with what has already been stated about sound being the principle creative factor of form from light. Electricity, like sound, is the passage of activity through a medium and involves the vibration of electrons. These little electrons are minding their own business, dancing away to their part of the resonance in the OM. They receive a burst of electricity, which is an influx of spiralling light waves which stimulates them and they become attracted to their opposite charge, the proton, joining together to become the neutron and creating the little neutrino baby, if you recall the description of this in Chapter 3 The Physics. This is happening on a large scale as many atoms are involved in a single cell and it could be that the sum total of the electron-proton interactions creates enough new neutrinos to create another new cell and so begins the life process. It is also important to say that neutrinos also travel in a spiralling motion as soon as they are created, in keeping with the fundamental behaviour of light.

Electricity can be seen as the activating aspect of the OM, and is perhaps the will aspect of consciousness, the creative factor of life itself because it is based on the principle of the positive and negative coming together to either cancel each other out if the charges are equal, or to create a third life (most likely if the charges are in the ratio of 1 – 1.618 (the golden ratio or phi). But all of life is based on the principle of the attraction of opposite charges, for example, the male entering the female. In the native American tradition they speak of Grandfather Sun and Grandmother Earth. Spirit, or light waves are the male aspect of consciousness, analogous to the sperm from the male, and vortices, or matter are the female aspect, analogous to the ovum from the female, so it is no accident that matter is generally referred to as feminine. Think of cars, boats etc. They are always referred to in the feminine gender. This is also where we get the name mater, or mother (from matter) and also why we call Earth, Mother Earth, or Mother Nature. And perhaps this is why we tend to refer to God in the masculine gender, because it is unseen, the spirit. But remember that matter and spirit are a reciprocal pair of the one whole Source.

Although we have described the analogy of masculine and feminine coupling in terms of lightwaves and vortices,

this principle exists on a higher level still. The lightwaves themselves being created by the continuous oscillation of equal and opposite charges, or positive and negative, or masculine and feminine, again different names for the same thing. This was first proposed by Michael Faraday to describe his field theory which were the lines of force emanating around two oppositely charged objects . The idea was later picked up by James Clerk Maxwell who described it with a set of four equations. He discovered that these lines of force were electromagnetic waves and that they had the same speed as light, concluding light must be an electromagnetic wave. It is now known that the faster the oscillation between equal and opposite or positive and negative charges, the higher the frequency produced, so a faster oscillation would produce the colour violet and a slower oscillation would produce the colour red, for example.. **The important point about all of this being that here we have some proof of the equal and opposite or masculine and feminine wholeness of light itself. If there were any need to prove the existence of the divine feminine as being exactly equal in importance to the divine masculine, this is it.**

So we can see that the electrical fields around living objects are a description of the morphic resonance fields. We can then say that the morphic resonance fields are electrical, what we sometimes call spirit or consciousness, or light. It's All Light anyway. The morphic resonance fields are the sound of the light, or slow light (but still faster than we can normally see). It seems we have different names for essentially the same thing and as I suggested in The Introduction, it may be easier to use the terms 'seen and unseen light' since they have less emotive associations for everyone. Following on from the fractal principle that there are harmonic levels of spiralling light waves then there must be harmonic levels of electricity. What we know as electricity is what we can bring it down to in the physical dimension. We actually only see the *effects* of electricity, not electricity itself. What we see is a precipitation of a force, just as matter is a precipitation of morphic resonance fields.

Magnetism is created by the spin action of electricity or the spin of charge from the positive and negative charges of electricity. An obvious model of this is the planet itself. Due to the pattern in the morphic resonance, the spiralling lightwaves in space stopped onto a point of spin, creating the vortex of matter which we call planet Earth. The lightwaves which are continually being pulled in to the spinning vortex are gravity but because they are electrical, they have positive and negative charges. The spinning of these charges creates what we call magnetism, which is none other than electrical attraction caused by spinning. If you can imagine an electrical lightwave spiralling along – the fact that it is spiralling means it will be creating magnetism. **This is what creates the electromagnetic wave, because there wouldn't be magnetism involved unless there was spiralling or spinning involved.** So we can say that magnetism is created from spinning electricity, which is further evidence that electromagnetic waves are spirals. When, due to resonance, the lightwave stops onto a point of spin, spinning in on itself, this creates the magnetic poles. As the out breath of Source, the OM, causes all light waves to travel in a spiralling fashion and some to spin on one point to take their place in the morphic resonance templates, then we can say that **sound is the cause of electromagnetism and magnetic poles.**

The scientist and author, Callum Coats, in the book *'Living Energies'*, reflects on Viktor Schauberger's theory that it is necessary to have the golden ratio for electricity to magnetism in order to create growth and that magnetism must be slightly in excess of electricism for evolution to proceed. The importance of this ratio is explained in Chapter 20 The Golden Ratio and Fractals. It is the ratio of 1 – 1.618 so the ratio of 1 – 1.618 is of electricity (1) to magnetism (1.618). This ratio seems to be important, perhaps fundamental, to the growth of form and is found throughout nature in, for example, some shells, animal horns, some seed patterns and the bones of the human body. For example, from the finger tip to the first knuckle then to the second knuckle is the golden ratio, then from the first knuckle to the second knuckle to the third knuckle is the golden ratio. The arm and leg bones are also similarly proportioned. It seems to be the ideal ratio to create expansion of growth and if you think of it, if something was in perfect balance there would be no need for it to grow neither would there be the stimulus. This seeming unbalance is a type of chaos and order does arise out of chaos. As with many things, for example, metallurgy. Atoms go through a chaotic stage prior to phase transformation. In the universe itself, order arose out of chaos. Perhaps chaos is another name for stimulation, or evolution. Perhaps electromagnetism is the method of evolution.

CHAPTER 17

GRAVITY AND THE STRONG AND WEAK NUCLEAR FORCES

As far as the laws of mathematics refer to reality, they are not certain, and as far as they are certain, they do not refer to reality.
Albert Einstein, Physicist and Mathematician (1879 - 1955)

All the forces of matter are musical harmonies within the morphic resonance fields. Gravity is nothing more than centripetal spinning, or in other words a vortex, in existence due to sound.

Einstein said that gravity, or acceleration, bends light but it is not known why. The conventional scientific explanation for why space is curved and planets orbit in circular pathways around larger bodies, like the sun, is because the planets and the Sun create a valley-like depression in space-time and the planets revolve around the curved depression, but again it is not understood why. The model used to show this exhibits a ball dropped onto a rubbery mat. Gravity pulls it downwards creating a dent, or depression, in the mat and we assume from this that on the life-size scale the dent is three-dimensionally created around the whole sphere. But this still leaves us with no explanation for what gravity is or why it creates the dent in space-time in every direction or why any orbiting planets have the particular orbit that they do. The reason they don't spiral inwards towards the planet they are orbiting is because of the law that says "in order to stay in orbit round the Earth (or other planetary body) an orbiting body has to achieve a velocity which will produce a centrifugal force which exactly balances the force of gravity", which as explained in Chapter 2 The Misunderstandings, is the force of levity. So it is levity which stops all the planets in this solar system from crashing into the Sun, according to this law.

What levity is –

Levity has been described earlier as complementary spinning *'achieving a centrifugal force which exactly balances the force of gravity'*. Centrifugal force creates a repulsion effect which can cause levitation. But spinning and counter-spinning are crucial to gravity and anti-gravity for another reason and this reason is electricity. Spinning creates electrical polarity which creates magnetism. Magnetism has both a positive and a negative pole. When two positive poles or two negative poles are brought together they repulse each other, thereby creating a levitational effect. So gravity and levity are intimately connected with electricity and magnetism. Spinning and counter-spinning are the key to gravity and levity.

The explanation of orbital pathways and the formation of planetary bodies -

Levity, however, doesn't explain the pattern of orbital pathways. While it is true that complementary spinning creates levity, I believe there is another explanation for orbital pathways. Again it is the morphic resonance field, or pattern. It has already been explained how the model of the atom is a fractal of the super cluster of galaxies. The atom's pattern contains a nucleus with orbiting 'shells' of electrons. This pattern follows on up to solar systems, where the Sun would be the nucleus and the planets the orbiting electrons, on up to the super clusters of galaxies where the pattern is repeated, consistent with fractals. The morphic pattern, or another way to say it, the sound, of the solar system has *defined* the shells. Each orbit or shell will be resonating to a different frequency and will be based on harmonics, consistent with morphic resonance theory. Johannes Kepler, the German astronomer who worked out the laws of planetary motion, discovered that the ratios between the planets' angular velocities were based on harmonic intervals and published this in his great work '*The Harmonies of the World*' in 1618.

This supports the morphic resonance of light theory in that it is all a great orchestral harmony, with the planets themselves being formed on the nodes of the pattern which is being produced by the sound. So morphic resonance can even account for why planets are the size, shape and position that they are and why they form at all. A node is a point of minimum disturbance (relative to its surroundings). To illustrate this look again at Figure 5. The higher the frequency the more complex the pattern, hence the more nodes. **Wherever the sound vibrates at the *fastest* frequency the *less* dense matter there is, so these are all the gaps where there is no sand on the plate (on a bigger scale this is space), wherever it vibrates at the *slower* frequency the *more* dense matter there is, so these are all the little clusters of sand on the plate (on a bigger scale these are planetary bodies) or the nodes in the pattern. This fact is consistent with the entire theory of the morphic resonance of light – that faster speeds (higher frequency) produce invisibility or less dense matter.** This can also be known as dark matter and is further explained in the next chapter. I have used the photograph of the cymatic experiment with sand on a plate to demonstrate this principle but it's important to realise that this process is not static. There is constant movement with continuous frequency. This is resonance – continuous sounding. Like in Figure 5, the cymatic experiments make patterns of matter from frequency and the matter can move in any motion whatsoever. It can spin, invert, undulate, anything at all and it is all produced by the particular frequency which is sounding. Likewise, if you can imagine the solar system, the galaxy, the entire cosmos as a great orchestral harmony – a continuously flowing, moving pattern due to the tune being played

Why space is curved –

Space is curved, or appears to be, because of the same reason for orbital pathways. It is part of the pattern. Remembering that light spirals, so it's curved anyway but spiralling is the inherent pattern of the cosmos. Also, the morphic resonance fields of the bodies of the suns and planets themselves, i.e. their template, is naturally curved because it's round, so it's like an extension of the shape of the planet. Space couldn't be anything *but* curved because of the fundamental vortexial motion of light. So, our solar system stays in place due to a pattern of frequencies inherent in the cosmic sound, which determines its shape, just like the atom. For an excellent description of this, with graphics to depict the remarkable coincidence of mathematics, geometry and the spiralling nature of the cosmos, see 'A Little Book of Coincidence' by John Martineau, published by Wooden Books.

An interesting aside is that the spiralling patterns and the geometry produced by planetary orbits, are reflected in many of the crop circle patterns which appear in our fields every summer, the source of which remains an enigma.

What gravity is -

The answer to what gravity is lies in the vortex, as described earlier. As the spiralling lightwaves of consciousness are flowing unheeded throughout cosmos, they respond to the sound (of the OM) taking the form of vortices, or sub-atomic particles (remember each sub-atomic particle is a particle in motion), which become atoms, which become planets; each an ascending size of vortex. Since we know that light is transformed into matter and that the sub-atomic particles and atoms emit light particles (photons) it means that each particle is dynamic: absorbing and radiating light. Spiralling consciousness/light waves constantly being made into particles to take their form within the pattern defined by the OM. The pattern is the morphic resonance field. The spiralling consciousness/light waves have no choice but to conform to the geometric pattern resonating. So there is a constant drawing into a spin of lightwaves to create the form. **This is gravity. It is a pulling-in, a stopping of lightwaves onto a point of spin due to sound. As explained above, the sound creates the node point which is why the lightwaves stop at that point. It would naturally spin because it is spiralling. A spiralling lightwave held on one point would naturally spin.** This also explains why gravity is greater for larger masses; why gravity is proportional to mass. A larger mass contains more sub-atomic particles, or vortices, hence the drawing in of more light waves and it is itself a larger vortex. The vortex is dynamic, as already described, with waves constantly being absorbed (gravity) and radiating (the life field or aura).

The life field, also known as the aura, is the radiation of light waves from each sub-atomic particle and can be seen around the human body, animals, plants etc. and can be photographed. This is also what creates the blue haze effect when the Earth is seen from space. It is the accumulation of the radiatory life fields of the planet and all life on it. Further explanation of this is given in Chapter 23.

The explanation of gravity being the dynamic absorption of lightwaves into vortices of matter also explains why falling objects lose some 'so-called' rest energy. The term 'rest energy' in physics refers to potential energy stored within matter, which of course is light. As has already been explained this light is spinning and vibrating to a particular frequency so 'rest energy' is a bit of a misnomer. Matter is never truly at rest. So the object, which is composed of dynamically spinning lightwaves, is pulled towards the larger vortex just like all the other lightwaves to take their place within the morphic field. A very small portion of the lightwaves of the object is absorbed by the larger vortex due to acceleration of lightwaves towards it; acceleration of the normal rate of absorption of lightwaves. Another way of saying this is, the falling object would retain its own morphic field (form) due to the law of resonance, so it would retain most of its lightwaves but the larger morphic field of the planet would command some absorption of all light waves in its vicinity because it is a more powerful field containing more vortices. It is the more dominant resonance. Gravity is lightwaves being compelled to take their form due to resonance. Or another way of saying it, gravity is the pulling in of lightwaves onto a point of spin due to sound. So, in essence, **sound is the cause of gravity.**

It is the same with the increased mass caused by travelling very fast. Mass and energy are equivalent terms. When an object is at rest it has a certain mass, or 'rest energy', which as I have described, is the appropriate portion of dynamic lightwaves. Experiments have shown that when an object is travelling very fast it appears to increase in mass, equivalent to the amount of force used to propel it. At close to the speed of light it takes ever more force or energy to move it at all. The object accumulates massive amounts of energy and becomes more difficult to propel.

What's happening here is that the extra mass is the accumulation of the force or energy (lightwaves) in each subatomic particle. Because the spin is linked to frequency and travelling faster means faster frequency, so the faster spinning particles are just keeping up with the faster frequency and, therefore, are able to absorb more light (mass).

In physics there is what is known as a 'light barrier' which means that no material object or particle can travel as fast as light because the accumulation of mass is too great. However, there is one particle which can travel at light speed and that is a photon (a light particle). The explanation is that it is because it has no 'rest mass', it is purely kinetic energy which when forced to stop, ceases to be, usually because it is absorbed, or more likely radiates away as a wave. But the photon is also a vortex created from a light wave. Photons simply display the quality of periodicity which makes them behave as waves, just like everything else in nature. They begin life as a wave, they can take the form of a vortex due to resonance, but they still behave as waves because waves are inherent in nature and are inherent in the morphic resonance templates. So light is essentially a wave but becomes a particle due to resonance and this is why light can be both a wave and a particle.

I would argue that it is possible for other particles to travel at light speed, remembering that they are also made of lightwaves. The greater mass accrued is simply an accumulation of lightwaves causing faster spinning, to keep up with the faster frequency (the photoelectric effect). From the particle's point of view, it's mass would feel the same and it's time would feel the same (faster spinning slows time) but for any observer left behind the particle would appear as extremely heavy and extremely slow. This is relativity, but that is exactly what it is. The reality is relative to the observer's experience. The frequency range of this dimension is relative to the speed of light at 300,000 kilometres per second. In the next higher harmonic dimension, say twice the speed of light, the frequency range will be faster, therefore, sub-atomic particles will be spinning faster (absorbing and emitting more light). When the object being propelled in this dimension reaches the speed of light the extra mass (accumulation of lightwaves) in the spinning sub-atomic particle will be in sync with the higher frequency (based on twice the speed of light) and it will appear to disappear from the physical dimension (much like the Starship Enterprise when it increases warp speed and disappears in a flash of light). There is often much truth to be found in the arts. Like the saying goes 'life imitates art'.

From the onlooker's point of view (in this dimension) the object when approaching the speed of light, appeared to have gained extreme mass and progressively travelled slower (faster travel slowing time), but to the object itself it would have seemed that it's mass was normal and progressively travelled faster. This is a perfect example of the merging of relativity and quantum theories. Both points of view are correct but are relative to the observer's consciousness, which is relative to the speed of light. So the fact that the experience is relative means that it is possible for particles to travel at light speed (and beyond).

What the strong nuclear force is –

The strong nuclear force is used to describe the force which holds the protons and neutrons together in the nucleus of an atom. Morphic resonance accounts for this force. See Figure 5 again to see how the nucleus can be part of the pattern of particular frequencies. The inner ring would be the nucleus, with the outer ring/rings being the paths of the orbiting electrons. The strong nuclear force is aptly named as it is indeed a very strong force, but this is of no more consequence than the weak nuclear force as it is an inherent sound in the cosmic Om and would remain as such as long as the Om is resonating and creating the universal pattern. The nucleus is part of the pattern. There is no need to describe a separate law to explain this force because it is inherent in the sound. If it had to be described as anything it would have to be described as a particular frequency within the overall frequency. **So sound is the cause of the strong nuclear force.**

What the weak nuclear force is –

The weak nuclear force is used to describe the process of the bonded electron and proton becoming a neutron. But as already discussed in the Introduction and in Chapter 3 The Physics, this can be more simply explained with morphic resonance. The weak nuclear force is also referred to as the force which holds the electrons in place, orbiting around the nucleus. There are 'shells' of orbiting electrons in an atom. Again morphic resonance accounts for this. The orbital paths are inherent in the sound, just like the orbital pathways of planets. When electrons jump shells it is because the frequency of the pattern has changed. The change can be caused by many things for example, an influx of electricity, remembering that atoms are not static, they are dynamic, constantly changing. So the whole pattern, structure and dynamism of the atom is inherent in the sound. **So sound is the cause of the weak nuclear force.** The electron and proton join temporarily, becoming a neutron. This is for the sake of creation of the neutrino which then radiates away and the neutron once more splits into the electron and proton. Very simple, analogous to biology.

So we have dealt with the strong nuclear force and weak nuclear force. Gravity is the dynamic force of light waves being pulled in to spin on an axis due to the morphic resonance pattern, to create matter. It is the inertia of the spin which continually draws in the light waves to form the vortex. Another word for gravity is implosion. Energy coming towards the centre, which is what the vortex spiral is. Another word to describe the energy radiating away as an aura is to call it explosion. As the dynamic life force radiates away it creates the life field or aura. It is really just radiation. So gravity really is just dynamic spin of particles which are compelled to take their place in the morphic resonance pattern. And orbital paths are the 'shells' defined by the harmonic frequencies of the pattern, analogous to the electron shells of the atom.

With morphic resonance we have a complete unified theory of the known forces of physics and an explanation of the life field or biophoton energy, also known as the aura. We can follow morphic resonance as the first cause all the way through the forces of physics: morphic resonance causes the vortex (spinning), the vortex causes gravity, which is the pulling in of lightwaves which are electrical, the spinning of which causes magnetism. So we can link gravity and magnetism as they are essentially the same thing, both caused by spinning lightwaves. The only difference is the electromagnetic wave is spinning whilst travelling, when it stops travelling and spins on one spot we call it gravity. The strong and weak nuclear forces are inherent in the pattern produced by morphic resonance, again due to spinning action which creates electromagnetism. Remembering that morphic resonance is sound which is creating the fundamental patterns or templates, so we can say that **Sound (which produces spinning action) is the unifying factor and the cause of the four main forces of physics.** However, in order to have a complete theory we need to include the lesser known forces in cosmology; that of dark matter and black holes.

CHAPTER 18

DARK MATTER AND BLACK HOLES

"There is radiance and glory in the darkness, could we but see, and to see, we only have to look. I beseech you to look."
Giovanni Boccaccio, Poet (14th century)

The invisible waves of consciousness is space, within which is the dark matter, which makes up 95 - 99% of the universe. These are the higher/faster dimensions, which includes the zero point field, the fifth force, funny energy, negative pressure, or quintessence as described by various scientists. It is also the cosmological variable, the 'n' in the equation $C = mc^n$.

In spiritual philosophy the physical is the densest level of reality for humanity– resonating to a low note, although it is possible that there are denser levels of reality since the electromagnetic spectrum extends in both directions with no finite end either way. the ascending levels of reality, however, in spiritual philosophy have been given different names, etheric, astral, causal, etc. and these correspond with the zero point field, or the other names mentioned above. They contain finer matter which is operating on a higher level of consciousness. It is resonating to a higher note. It is in existence relative to a harmonic multiple of the speed of light. The important word is <u>relative</u>. The physical law of science says everything in this world is <u>relative</u> to the speed of light – which it is. Einstein said when we approach the speed of light time slows down – which it does. So the physical laws are relative to the physical world.

So my argument is there are speeds of light which are non-physical, a portion of which may be vibrating at a slower harmonic speed of light but evidence points to there being many more faster harmonic speeds of light, vibrating at ever increasing frequencies, which enable a higher or greater consciousness. The argument for which is given in Chapter 15 Proof of Speeds Faster Than Light and which can be seen in Figure 5, which showed that increased frequency creates increased information. These non-physical waves of light/consciousness travelling in harmonic multiples of the speed of light are the <u>dark matter</u> because wherever you have waves you can have particles. The dark matter is finer or subtle matter. This is the missing matter in the universe which science has been looking for. The dark matter is <u>light</u> waves but invisible to us tuned into this <u>physical</u> speed of light. The dark matter is LIGHT. Dimmed to our view because of the speed of our consciousness. As described in earlier chapters, the photo-electric effect shows us that the energy of the particle of light determines the frequency. The energy of the particle of light is determined by the amount of spiralling light waves entering the vortex/particle, which determines the rate of spin. If you remember, the rate of spin determines the time factor which determines the frequency that the particle can live within. When the frequency is very high, physical matter including light, appears to disappear. So we can have particles of matter vibrating at very high frequency beyond the speed of light. This is what we call dark matter, though that just means invisible. If your consciousness was tuned to that very high frequency the matter would be visible.

Black holes are also light that we cannot see, as well as being spinning vortices. They appear to be black but, like space, that just means that we cannot see what's there because we are not tuned in to that frequency. At the beginning of my research I believed that black holes were vortex entrances to higher/faster dimensions. Much like the tunnel of light that people see when they die. This could still be the case. However, they could just as easily be vortex entrances to lower/slower dimensions. This is due to the discovery that they have been found to resonate at 57 octaves below middle C. If you remember that 40 octaves above middle C gives us our colour spectrum, the colour blue in fact, and the whole octave around middle C, from lower F to upper F gives us our colour spectrum from red to magenta (or ultraviolet). Using the same principle, 57 octaves below middle C would take us well beyond the colour red out to the far end of the electro-magnetic spectrum at the radiowave end. It could even be outwith our spectrum in a slower dimension but what makes me think it is probably within our spectrum is because we can see it. We cannot see its light or colour, because we have a very narrow vision range, but we can see the vortex. The black hole is a spiralling vortex. Everything is drawn into it just as the spiralling waves of consciousness are drawn into each little vortex. The same principle applies. It is light spinning and vibrating outwith our range of vision. It could be a harmonic multiple of the speed of light but in this case it could be a lower harmonic. A subset of the speed of light rather than a superset. We still wouldn't see the light for the same reason that we cannot see microwaves or radiowaves (they are outwith our range of vision) just as X rays and gamma rays are outwith our range of vision at the opposite end of the spectrum.

So black holes, I believe, are vortex gateways to other dimensions. Other vortex gateways are the tunnels of light that people see at the point of death or during a near-death experience which are vibrating at a higher frequency than the known speed of light, a superset of the speed of light. This tunnel of light experience is a universal one regardless of culture or religion. At the point of death the person becomes aware of a light which becomes a tunnel and they feel drawn towards it. This is consistent with a vortex which draws all towards the centre. The fact that they see it as light means that their consciousness will already be operating at a faster speed due to their being outwith the confines of the physical body. It is possible, on occasion, or with particular training, to see this light even while within the body, while still alive, but at the point of death it is automatic, without effort. The consciousness is fully released from the speed restrictions of the physical body.

The reason there is so much dark matter is because there are many levels of existence beyond the physical. The physical world is a very tiny aspect of the whole of reality. In this dimension we are blind to 95-99% of reality unless we can change the speed of our consciousness. It is by this method that the energy in space can be tapped. This is basis of healing and zero point technology. In healing, the consciousness is tuned to a higher or faster reality and allowed to flow through the healer to the patient. And healers will tell you that it is light that they are attuning to and light which they see being transmitted to their patients. They also feel this light is love, that they are synonymous terms. The ability of the healer to tune-in to the many different levels of consciousness available is one of the factors which reflects the outcome of the healing. In zero point technology it is also being tapped but for a different use and probably from a lower harmonic than for healing but still at a faster speed than our known speed of light.

So in the last chapter it was discovered that gravity, electromagnetism and the strong and weak nuclear forces can be explained and unified by morphic resonance. We have already reconciled relativity with quantum theory, they are just opposite ends of a spectrum of speeds of light and we can now see that dark matter and black holes are just unseen light. We have already said that light is consciousness and contains within it sound, or resonance, which means continuous sound, so the equation given in Chapter 11 Consciousness is the Key, still stands. $C = mc^n$. Consciousness (light and sound) equals matter times the speed of light in harmonic multiples (creating matter and subtle matter). But matter is spinning light – so – IT'S ALL LIGHT. A complete theory of physics. However, if you remember earlier it was explained that the original meaning of physics comes from the ancient Greek word 'physis' which means the understanding of the fundamental, eternal essence of all things and all life, then to have a complete theory or a truly unified theory, what we call the spiritual or unseen realities would have to be included. Not just the unseen reality (dark matter) as described in $C = mc^n$, which is simply a physics theory or mathematical concept, but the unseen reality which is very often 'seen' and is a natural part of human experience the world over, which we often call spiritual phenomena. Any true scientist would not overlook overwhelming evidence and the evidence for spiritual phenomena is truly overwhelming.

CHAPTER 19

SEEING WITH THE HEART

> *"It is only with the heart that one can see rightly; what is essential is invisible to the eye."*
> Antoine de Saint Exupery, Aviator and Writer (1900 - 1944)

When one can attune the consciousness to a higher speed it becomes possible to access realms where time has slowed down and doesn't exist, or the past, present and future co-exist, which is, of course, relativity. This accessing of timeless realms is what makes possible such abilities as clairsentience, clairaudience and clairvoyance, meaning clear sensing, clear hearing and clear seeing, respectively. This is why it is possible to see into the future, for those who have clairvoyance, because they are 'tuned' to a faster speed of consciousness. They are seeing the model that relativity/determinism speaks of, past, present, future all being one or a state of no time, or another way of saying it, they are seeing the spiritual realms, the unseen light. The word spiritual is only used to denote that which is non-physical or unseen but is another example of dualistic thinking, matter vs. spiritual. It is a tool of language to explain the limitations of our understanding or experience, but in fact, as all Life is dynamic and interconnected, in a sense there is no differentiation between matter and spirit. As explained earlier, we have this dualistic concept of life due to the level of our consciousness.

A necessary part of attuning the consciousness to a higher speed is by opening the heart centre or 'chakra' as it is known in the healing arts, which is the Sanskrit word for 'spinning wheel' because this is what it looks like. It is simply another vortex with energy flowing in and out. To open the heart centre is to allow the dynamic uninterrupted flow of love and compassion. It is primarily through this centre that we feel love and compassion. In Chinese thought it is believed that it is possible to comprehend the entire universe through the vehicle of the heart. As discussed in Chapter 5 Holistic Consciousness, it is necessary to have a harmony of expression from both sides of the brain, but it is also necessary to 'open the heart' and new findings seem to suggest that it seems to be the heart centre which is crucial to the evolutionary process, to the achieving of enlightenment or omnipresence, and this is explained further in the next chapter.

In his book *The Healing Energies of Light*, Dr. Roger Coghill describes how, in the laboratory, they discovered that the viability of white blood cells, crucial to our immune system, is profoundly affected by what he calls the 'endogenous electric field' of their owner and that this field originates in the brain and heart. The 'endogenous electric field' is what I have referred to as the aura, or biophoton field – again like so many other things, different names for the same thing, but it's not surprising that he describes it as an electric field, if you remember my earlier description of electricity. Roger Coghill's work is not only further proof of the aura but also confirms experiments described in Chapter 22 The Heart and Quantum Concepts which show the effects of a coherent functioning heart. One way of creating a coherent functioning heart centre is through music. As explained earlier, finding the correct sound brings us back into resonance. As we are composed of a morphic resonance field or template then being in resonance will keep us healthy and functioning well, whereas dissonance will create illness or disease.

As explained, the morphic resonance fields are fractal, which means there are fields within fields, or dimensions within dimensions and another example of the existence of the morphic resonance fields is what is known as the harmony of the spheres. This is a phenomena reported by people who have had near death experiences and often heard during profound spiritual experiences, and taught by great teachers and esoteric schools throughout history. It is the harmony made up of the higher tones of the ascending levels of consciousness, in effect, the harmony of the OM. Strong evidence of this harmony and of the dynamic and interconnected nature of life throughout the realms is overtone chanting, now taught in workshops around the world. When a certain note is intoned and held, a higher note can be heard simultaneously. It usually sounds like several higher notes; a mini harmony, quite beautiful. This is convincing evidence of our dynamism as part of the music of the spheres. We are dynamically interactive with other realms of existence. We usually say higher realms but it may be more correct to say faster realms however it may take some time to integrate this saying into the language to have the same meaning for most people as higher realms. Especially when we talk of being 'lifted' by a spiritual or healing experience, or 'lifting our thoughts', the 'higher mind', 'raising our consciousness' etc. and people tend to think of a physically higher place. These faster realms however, are really not so much 'up there' as within us because all the morphic resonance fields are fractals. So the same potential is within every fractal. In effect, we *are* the harmony. The potentiality is inherent in every light wave and hence every particle as ultimately it all comes from the same source. 'The kingdom of heaven is within us' as Jesus said. Yet few really know what this means. The task for us as humans is to realise (literally, to make real) that potential. As above , so below. The fractal principle. The hologram principle.

We are a microcosm of the macrocosm. All Life is dynamic and interconnected. So we have unlimited potential within our consciousness. The question is 'how do we access it?

Many teachers throughout the ages have told us the method. Love one another. Masaru Emoto's work with water crystals also proves this teaching. As he demonstrated, love and compassion create harmonic and coherent water crystals and since life on Earth is based on water it seems to be a very good law. Also, this is reflected in Jacques Benveniste's discovery that water is the medium by which frequencies are transmitted throughout the body, and Kunio Yasue and Stuart Hameroff's discovery that coherent, or ordered, water creates coherent photons. So it seems that Love is the law. This creates coherent water, which creates coherent photons. The importance of this law is explained in the next chapter on the Golden Ratio and Fractals. This chapter explains the physics of what happens to each vortex when we love.

CHAPTER 20

THE GOLDEN RATIO AND FRACTALS

> *"The human heart is humanity's great secret weapon. Once awakened, it contains more power than the atomic bomb."*
> Teilhard de Chardin, Scientific philosopher (1881 - 1955)

In studies using superconducting quantum interference devices, magnetocardiograms and magnetoencephologrums which measure magnetic fields outside the body, it has been found that the heart generates over 50,000 femtoteslas (a measure of electromagnetic fields, or EMFs) compared to less than 10 femtoteslas generated by the brain. So we can see that the heart's electromagnetic field is 5,000 times more powerful than the brain's electromagnetic field [1].

When we love, our heart rhythms come into an alignment. In other words, if a person's heart is hooked up to an ECG monitor and they allow themselves to feel love and compassion, with each output the waves become more coherent and eventually embed within each other. This was demonstrated using real time measurements at a seminar by Dr. Dan Winter. As we know in physics when waves come into phase, or embed, this creates coherence. In other words, all the waves are focussed together and not scattered so energy accumulation can be enormous. In the ECG experiment, when the person feels unbalanced in any way or not feeling love and compassion, the pattern destabilises and becomes incoherent. The same as we see happening with Masaru Emoto's water crystals. So what we are really talking about here is coherent wave patterns which creates enormous energy potential. Dan Winter's work has been replicated by the Heartmath Institute in Boulder Creek, California. [2]. Professor William Tiller, Department of Materials Science at Stanford University has spent decades investigating intention and intuition and the relevance of coherence in these processes. In speaking about coherent wave patterns and the potential this creates, he uses the example of a 100 watt light bulb. As the photons are emitted from the bulb they cancel each other out in a process known as 'destructive interference', in other words, the waves cancel each other out so you never see the full potential of the light bulb. But if you could somehow orchestrate the movement of the photons to come out 'in phase', or coherent, the energy density from the bulb would be between 1,000 and 1 million times that of the sun.! If love is indeed expressed by coherent light waves then what does this tell us about human potential? !

Dan Winter has written and lectured on the significance of the golden ratio and other aspects of sacred geometry. He has discovered that the only way that waves can enter into each other non-destructively *in a spiral* is if it is a golden ratio spiral. The golden ratio being 1 – 1.618, expressed as phi and written as φ. As described in the chapter on electromagnetism, this ratio may be necessary to create life. It certainly seems to be prevalent throughout nature, and, if Dan Winter is correct, it may be the ratio necessary for waves to attain coherence in a spiral. If this is correct then the golden ratio spiral would be the ultimate model or design of the OM, since the spiral is the fundamental design of light.

The golden ratio is a ratio which is pleasing to the eye and it is often found in ancient architecture and paintings. Leonardo da Vinci for example, used this ratio in many of his paintings. Many artists and sages from the past knew the significance of this ratio and knew its significance in the natural world and in universal principles. Plato was one of those. He carried on work which was started by the Pythagoreans of ancient Greece, into the form and structure of matter using the principles of sacred geometry, particularly the polyhedra, or Platonic solids as we call them today. The platonic solids are tetrahedron (4 triangular faces), cube (6 square faces), octahedron (8 triangular faces), dodecahedron (12 pentagonal faces) and icosahedron (20 triangular faces) (See Figure 6). They are unique in the sense that they are the only solid shapes which have been found to have the following properties: a) all the faces of each solid are identical and equilateral and b) each solid can be circumscribed by a sphere where all its vertices lie on the sphere.

The golden ratio is essential to the structure of the dodecahedron and icosahedron and all of the platonic solids can comfortably fit within them. So the golden ratio is a crucial ratio to the structure of form. Two very good books describing the geometry of this are *'A Little Book of Coincidence'* by John Martineau, published by Wooden Books (mentioned earlier in the text) and *'The Golden Ratio'* by Mario Livio, published by Headline Book Publishing.

Another intriguing aspect of the golden ratio is that when it is drawn as a rectangle it is the only rectangle which, when cutting a square from it, produces a similar rectangle. The dimensions of the smaller rectangle leftover are smaller in proportion to the larger rectangle by precisely phi, the golden ratio. This dimensional relationship continues endlessly in scale downwards or upwards, which makes it a fractal.

A fractal is when similar patterns recur at progressively smaller scales e.g. on a pine tree, the tip of the needle is the same shape as the branch which is the same shape as the tree itself. The pattern is also true of the large scale. Several decades ago Benoit

Mandelbrot, who was educated at Cambridge and worked for IBM Research Centre, discovered the fractal nature of galaxies and this has been confirmed with the newest telescopes which can look deep into space to a distance of 650 million light years. Clusters of stars become clusters of galaxies, then super-clusters of galaxies and so on forever increasing in scale; the fractal principle. Mandelbrot described this principle in a geometrical form which has since become known as the Mandelbrot Set. Each small piece of the drawing is a replication of the whole and it can continue down or up in scale to infinity. So the fractal is a universal principle and has been alluded to in many ancient scriptures, such as 'as above, so below' but we can see the evidence of it all around us.

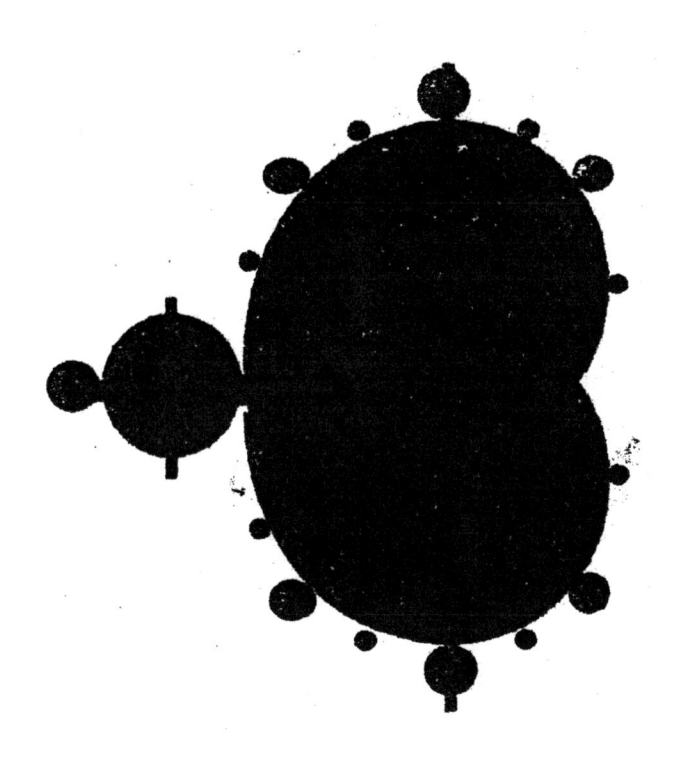

FIGURE 9 - THE MANDELBROT SET

Going back to the golden ratio rectangle, if you connect the points where each rectangle is divided into a square, you obtain a logarithmic spiral, showing the intimate connection between the logarithmic spiral and the golden ratio. There are many golden ratio spirals that can be created by varying the degree of angle between adjacent radii (from the central point) but the one in Figure 10 is the more common 90° phi spiral and is the one often seen in nature, for example, nautilus shells. However, phi at lower angles, such as 15°, 30° or 45° produces leaf shapes, so again this is another way that phi is prevalent throughout nature. .

The *logarithmic spiral* is unique in that it keeps the same shape regardless of how small or large it becomes. This is known as **self-similarity** – which like a **fractal**, retains the same pattern at progressively smaller scales and like a **hologram**, contains the pattern of the whole within the part. Good descriptions of wholeness or infinity or 'made in the image of the Creator'. These descriptions are also true of the *golden ratio spiral*. A good way of describing self-similar, fractal holograms would be infinity. Infinity is also a description of the calculation of the golden ratio, or phi.

It is known as one of the Divine Proportions and, like pi, is a never-ending number. It has been calculated to 1: 1.6180339887 … etc.etc. unending. Pi, on the other hand, describes the ratio of the circumference of any circle to its diameter and is calculated at 3.141592 …….etc.etc. unending. Both have no end to the sequence of numbers after the decimal point. Good descriptions of infinity or Wholeness. Now one of phi's structures is the spiral, (spiralling lightwaves creating form) and it has been described as the OM. *The breath of the light.* Pi represents an infinite size of circle, The circle is universally recognised as a symbol of God. *The light before it breathes.* The circle (pi) breathes forth the spiral (phi) to once again become the circle (pi, or the vortex particle) 'as above, so below' 'made in the image of the Creator'. It's also interesting that 'phi' is embedded in mor'phi'c (the phi spiral is embedded in form) and pi is found in p(h)i. - *it's All Light*. The use of words is no accident.!

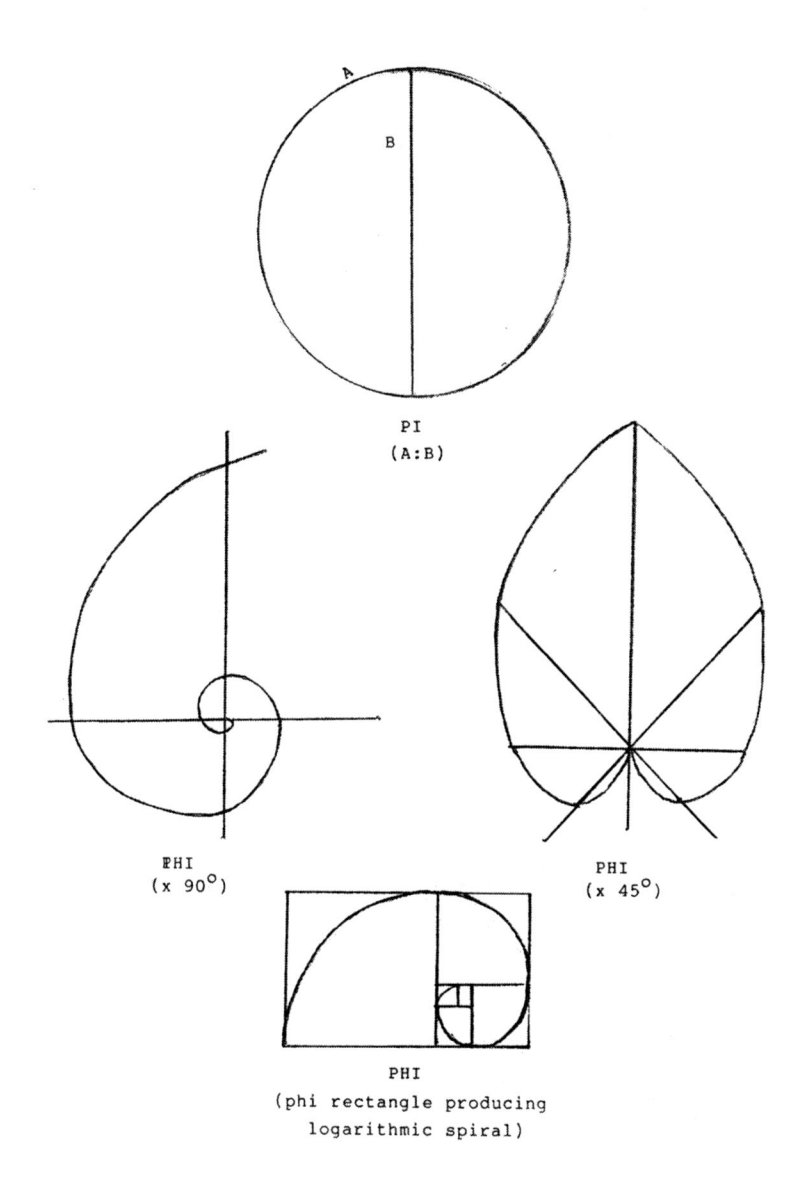

FIGURE 10 - PI AND PHI (EXAMPLES OF A GOLDEN RATIO SPIRAL AND LEAF SHAPE AND A LOGARITHMIC SPIRAL PRODUCED BY A GOLDEN RATIO RECTANGLE)

Nature seems to have found a way to design life out of the spiral (the universe, uni – one, verse = curve), whether a simple spiral, golden ratio or logarithmic, the latter being produced from golden ratio measurements Since feeling love and compassion cause spiralling lightwaves to embed or become coherent, then a way of describing becoming centred could be –

Coherent heart rhythms embed within a (golden ratio) spiral, or
Coherent heart rhythms are created by Love, so
Love embeds within a (golden ratio) spiral, or
A (golden ratio) spiral is the pathway to and from Source, so
Love embeds with the pathway to and from Source, which is the same as saying,
Love becomes One with Source

Could this be why all the great teachers who have visited the Earth have specified that love is the key? . Love and compassion have been the teaching of these great ones throughout history. Perhaps we now know why. It is the key to returning to the Source; to becoming One with all of Life; to become centred.

Perhaps this is the 'pull' that is felt towards finding meaning in life; the striving to reach centredness which is a really good description of peace. We all know when we feel unbalanced or off-centre, we even use this type of language. Like the examples given previously, I believe the use of words is no accident, for example, we talk of a building or other structure being 'sound' or talk of someone being of unsound mind. Sound remember is resonance and so we are talking of being in resonance either harmonically (coherent) or unharmonically (incoherent). The incoherent resonance creates an unbalanced structure. In this light we can make sense of some ancient mysteries, for example, the story of the walls of Jericho being destroyed by the blast of a trumpet and we know that glass can be shattered by certain sounds. Even in the medical profession sound is being used, called lithotripsy, to break up kidney stones. I also believe that certain sounds can plasticise particles then reassemble them, simply by altering the morphic resonant field temporarily (through changing the sound), and this I believe, is how the pyramids were built, which would explain why the stones fit together so perfectly and why some stones within the pyramids could not physically have been placed there as solid objects.

It has been described above how all life is spiralling waves and that love creates coherent or embedded waves then love is obviously the key to becoming coherent or centred, in the spiral. Remember the fractal nature of the universe, there are spirals within spirals within spirals. From sub-atmomic particles, through DNA, on up to the galaxies – all spirals. To be 'at One' would be to be embedded within the spirals, or centred. Centreing is also known as 'going within' to reach that place of enlightenment or realisation of our essential nature, what some may call our divinity. So Love seems to be the master key for access because it creates embedding or coherence and coherence ultimately creates non-locality, or omnipresence. So this describes the reason for the teaching of 'Love one another'. There are very, very few individuals on Earth who have mastered this. Most people can just about manage to love their children, spouse and some other relatives, perhaps close friends but some of these are due to soul connections, familiarity and ties. Love one another means everyone and everything, including Nature, and to be in a state of complete absence of fear and separation. To have mastered Love means to be in a state where the experience is one of complete and total interconnectedness with everything, to see that there is only one living force moving through everything and to experience intimately that to harm anything or anyone is a direct harm to oneself. Not only this, but to have true empathy; to feel the joy or pain of another as if it were your own (which in a sense, it is) Love does not include pity, sentimentality, possessiveness, lust or familiarity, all of which are often mistaken for Love. It is obvious that very few people manage to master Love within one lifetime. It is the result of a long evolutionary journey. This is the reason for reincarnation.

CHAPTER 21

REINCARNATION OR REBIRTH

> *"Every soul comes into this world strengthened by the victories or weakened by the defeats of its previous life. Its place in this world as a vessel appointed to honor or dishonor is determined by its previous merits or demerits. Its work in this world determines its place in the world which is to follow this ......"*
>
> Origen (one of the Church's founders) (185-254 A.D.) (from 'Principiis')

The early Church taught the doctrine of reincarnation and it is frequently referred to by the great philosophers of the early Greek period, Pythagoras, Plato and Socrates. It was a surviving belief until around 402 AD when Pope Theophilus crusaded against Origen and gradually the doctrine began to fall into disrepute with the church. However, it wasn't until the Ecumenical Council in 553 AD, also known as the Second Council of Constantinople, convened by Emperor Justinian, that the teaching was removed from the Bible. Justinian believed it to be against Catholic doctrine. It has survived, however, in the Eastern hemisphere and there are many today in the Western world who still believe it is an exclusively Eastern concept who know nothing of its original inclusion in the Bible or that it's teaching extends to pre-Christianity.

There have been several studies into reincarnation where people have remembered events of previous lives and have even been able to accurately recall people and places relating to a former life. One notable example of this is the case of Jenny Cockell. Jenny was born in England in the 1950s and from the age of 3 talked of her life as 'Mary' and her children and where she lived in Ireland. As she grew older she recalled feelings of concern over her children's welfare. She began to draw maps and diagrams of her former home and eventually in her 30's she tracked down her 'family' of those days. By now she was married and had a family in this life. She managed to trace all her children from that life, most of whom were now in their 70's. She knew everything about them and talked of their life in Ireland, of events and even family secrets that no-one else knew. Some of them have accepted that Jenny was their mother from a previous life, some just remain open to the possibility. Jenny's concern for her children was due to the fact that she had died in extreme anxiety as a young mother leaving 7 children behind with a father who was frequently absent. The children had been placed in orphanages after she died but the feeling of anxiety and concern carried over into her next life (this present one). Jenny's story is documented in the book *'Beyond Time and Death'*.

Other cases where people remember previous lives also indicate that the person or soul retains the strongest emotion at the point of death and it is carried over into the next life. This is one reason why spiritual healing can be so helpful to someone even at the point of death, because it helps them to make the transition peacefully and without fear, reincarnating into the next Earth life in the same state. Not enough is known about time intervals between reincarnation and this aspect seems to be pure speculation, but it seems obvious that for a person to die in peace rather than fear can only be a good thing. This healing benefit is something I have observed from my own experiences. A dying person received spiritual healing and was able to revive enough to, not only leave their bed, shower, dress and enjoy a family day, but crucially included a final cathartic conversation with the family which brought a healing to all of them. We can't know the soul's path and sometimes a healing can be physical and the person continues to live physically, sometimes they die physically but are spiritually healed. By that I mean they achieve a peaceful state which is the natural state of the soul. It is always present but physical, emotional and mental demands, beliefs and disturbances can by so overwhelming and consistent that they literally shroud the natural state. This is the state of most of humanity because of the way the world is. The word 'spiritual' in spiritual healing simply refers to the unseen reality where the energy comes from, the unseen light, where the peaceful state resides. It in no way implies a religious belief system, although religious people may also practice spiritual healing just like anyone else but it is not the preserve of religion. It is spiritual or universal and has been around for as long as people have been around. It's often called 'natural healing' and it is very natural to reach out to help another, to offer compassion. The healing involves invoking the unseen light (although it can be seen by some individuals) and channelling it to the person requesting healing, usually being directed through the hands, although it can also be done by the mind alone and directed anywhere in the world. This is known as distant healing and it works due to non-locality, as described in previous chapters. Light (or energy if it helps to use the conventional word) is everywhere and can be directed with intention. This is explained more fully in the next chapter which documents experiments which have proved this. There are many different methods and types of healing available these days, most of which are ancient and have simply been revived, but it is an exciting time in the healing arts today as more and more is being discovered about the energetics of the human system and many of the discoveries about reincarnation are coming from this.

There are even cases where some people (usually children) remember being killed in a particular manner and they have an unexplained mark on their body which is where they received the fatal wound in a previous life. These have been documented in the late Dr. Ian Stevenson's study called *"Reincarnation: Field Studies and Theoretical Issues"* published in Handbook of Parapsychology.
[1] Dr. Stevenson was educated at St. Andrew's University in Scotland, McGill University in Montreal and at the age of 38 became professor and chairman of the Department of Psychiatry at the University of Virginia. He spent the earlier part of his career studying the effects of stress and emotions on the body, then investigation into the psychiatric effects of psychedelic drugs which led to research on psychic phenomena, including near death experiences, extra-sensory perception, clairvoyance etc. But the last 35 years of his career was spent on investigating people, mostly children, who experienced memories of a previous life. He studied almost 3,000 cases of this nature, more than 200 of which included birth marks which corresponded with the fatal wound from the previous life. Dr. Stevenson's voluminous and detailed research, more than 300 publications, have been largely ignored by mainstream science. Again, we have the old problem of threat to world view by orthodox scientists, probably because it's dealing with the non-physical. This is the same as what happened to Howard, Abrams, Benveniste, O'Leary and others mentioned previously. Subjects like reincarnation, healing etc. are voodoo to many scientists but to ignore such a body of evidence is very unscientific. While many scientists do not want to play science, it seems, thankfully some do.

The strong emotions carried over from the previous life would also explain so much about unexplained fears etc. in very young children, though it's not the only possible explanation. Reincarnation also explains unusual abilities, untaught skills, prodigies and many other unexplainables because the same soul is accumulating experience with each lifetime. Each soul is part of an original morphic field, the original source. This remains and if you recall, all memory fields are contained within the morphic resonance so the soul has the accumulation of all its memories. Very young children often speak of a previous life, before the idea gets conditioned out of them and eventually they forget, most often by the age of 5. This is remarkably common. Reincarnation is also the working out of the law of action and reaction in physics, also known as karma in spiritual terminology. It is a law in physics which means that for every action there must be a reaction. This law goes hand in hand with reincarnation because it is the process by which all karma is eventually worked out; the reaction. It clearly cannot all be dealt with in one lifetime. It has already been described how Love is the key to enlightenment or returning to Source. Returning to Source. Returning to Source means liberation from form, in other words we don't have to come back again to be born into another body. We only have to keep doing that until all karma is cleared and Love has been mastered.

We have already seen how consciousness, in the form of thought, and words affect matter, and we know that action affects matter, so in order to clear all negative karma, every thought would have to be Loving, as well as words and actions, but thoughts are more difficult to control. Thought is an extremely powerful force and it is very important because words and actions arise from thought. Thought arises from consciousness but consciousness itself is much more than thought.

In *'A Treatise on Cosmic Fire'* published in 1925 by Alice A. Bailey, karma is defined as "the force generated by a human centre to act on the exterior world, and the reactionary influence that is in turn generated from the exterior world to act on him may be called karmic influence and the visible result that is produced by this influence under proper conditions may be called karmic fruit.". So resolving karma is obviously important to mastering Love. In other words, we have to keep returning until we have mastered Love because any thought, word or action which is not loving will create incoherence. This is probably why every great spiritual teacher has taught the law of karma, as well as the law of Love. Some examples of this law of karma are as follows: - "As you sow so shall you reap" – Christian. "Hurt not others in ways that you yourself would find hurtful" – Buddhist. "Regard your neighbour's gain as your own gain and your neighbour's loss as your own loss" – Taoist. "What is hateful to you, do not to your fellow man. That is the entire Law, all the rest is commentary" –Jewish. "This is the sum of duty: Do naught unto others which would cause you pain if done to you" – Brahman. "Surely it is the maxim of loving kindness: Do not unto others what you would not have done unto you" – Confucian. "That nature alone is good which refrains from doing unto another whatsoever is not good for itself" – Zoroastrian. "No one of you is a believer until he desires for his brother that which he desires for himself" – Islamic.

It seems to be Christians who have the hardest time with the concept of reincarnation. Yet if one thinks of this logically, there is no satisfactory explanation why a loving, benevolent God would tell us we must accept Jesus Christ as our lord or we will not go to heaven but at the same time put some of us in life situations, say born into a non-Christian country and perhaps die young, where the chance to know about Jesus never arises and you don't get another chance. You have one life only. The way out of this is that on Judgement Day you will be resurrected and then get the chance to accept Jesus to be saved. But this would hardly be a choice considering the alternative and would the individual have learned any valuable lessons? This approach is also judging the whole life of the individual as having been misled, misguided or 'not chosen'. The reasoning just doesn't follow and allows little space for inclusiveness, which was central to Jesus' teaching. It is also placing human limitations of love onto God. Also the many millions of people on the planet who practice a different religion and who achieve a certain oneness with the divine, are they wrong? misguided? not chosen? It can only be ignorance or arrogance to believe that millions of people are so misguided or not chosen. The church, while arguably keeping alive the teachings of Jesus, have alongwith other religions been just as steeped in its dogma as science has in its.

Even if one doesn't believe it is necessary to accept Jesus (the personality) to enter heaven, but as argued above, it is necessary to master Love (which he obviously had mastered and which he demonstrated) in order to return to Source, then it is obvious that it doesn't happen in one lifetime. Think about how many people you know who have mastered Love. It is extremely rare. Most of us do not know one single person who has mastered Love, though we may know people who practice it more than others. It requires total unconditionality and the mastery of the senses and in particular, the mastery over the ego which really means the sublimation of the ego. This is necessary because the ego masks the true authentic self, or divine self, or soul as it is sometimes called. The ego refers to the entire personality including all its attachments, prejudices and conditioning, particularly conditioning because true Love is unconditional. And attachments include all the things related to our self-image, such as nationality, religion, politics, financial status, professional status, etc. all of which are divisive if there is any kind of attachment. If there is division against any part of humanity then the state of unconditional Love cannot be achieved. The ego is really just mental and emotional constructs (the attachments, conditioning etc.) which are built up from birth, some of which are carried over from previous lives, and act like bars on a cage with the soul, embodying transcendental Love trapped inside. Through the evolutionary process, as the karma is cleared, Love is learned (or perhaps it may be more correct to say Love is remembered) and through education the ego is dissolved, this enables the bars on the cage to gradually dissolve and the soul can more and more reveal itself through the individual. The last step on the path being a totally soul-infused individual. This metaphor may be helpful to understand why reincarnation is necessary. Karma is often thought of in negative terms but it's just a reactive energy, and it's important to realise that because we are one humanity as part of one life, this means we are also part of group karmas and national karmas. There are many factors that come into play in the life of one individual. While it's true that, what we focus attention on we can bring into our life, it's not the whole picture.

The law of attraction is being talked a lot about today and of course karma seems to work like a law of attraction but it wouldn't be true to say they are the same thing. The Law of Attraction is described in the esoteric philosophy, also known as The Ageless Wisdom teachings which were channelled through various individuals by a Tibetan sage. By that I mean they were imprinted mind to mind via telepathy, which as we know, is possible due to non-locality of consciousness. In, for example, *'A Treatise on Cosmic Fire'* mentioned earlier, which was published in 1925, Alice Bailey describes the law of attraction as one of many laws. For example, there are eleven subsidiary laws under this law, namely, 1) the law of chemical affinity, 2) the law of progress, 3) the law of sex, 4) the law of magnetism, 5) the law of radiation, 6) the law of the lotus, 7) the law of colour, 8) the law of gravitation, 9) the law of planetary affinity, 10) the law of solar union, 11) the law of the schools. Then under what she calls 'Motion on the Plane of Mind', there are four laws, 1) the law of expansion, 2) the law of monadic return, 3) the law of solar evolution, 4) the law of radiation.

So it seems simplistic to say that what we focus attention on we bring into our lives, considering all the other factors involved, although it is true as part of the bigger picture. The hardest part of learning focus, however, is in identifying what it is that we are focusing attention on. It takes absolute honesty with ourselves, something we're not very good at but it also takes a certain awareness to realise that we have been conditioned from birth and that the self-identity and attachments we have are all due to conditioning and that this is what makes up the ego. It's not just big-headedness, which is what most people mean when they talk of ego. It's the whole personal identity with all its attachments and conditioning and is a kind of prison. It keeps us from knowing our true nature – which is the soul, the hologram, light, love, all the same thing. True love is an ego-less or self-less state and is achieved through the evolutionary process. This is what it means to be 'At One with everything', or 'we are All One', as taught in every spiritual tradition.

It is important to say something here about the way the law of attraction is sometimes presented, as there is a great misunderstanding amongst some in the new age community who are causing guilt on a scale comparable with the Catholic church, by saying that everything that happens to you is your own fault, that you created it yourself, that if a tragedy befalls you, you somehow must have created it or deserve it as some kind of payback. This can come across as incredibly self-righteous and cruel and is a misrepresentation of a great law. The law is there to teach us the power of our thoughts and the potential of our creativity. It is not a punishment. The personality of course, would rather not have the difficulties but remember there are other factors at work such as group karma and national karma and karma going back many lifetimes, etc. As well as this the soul may bring in these difficulties in order to achieve mastery, mastery over the senses and over the mind, your own mind and senses that is, and not someone else's. So to see someone who is obviously struggling with many life problems as someone who is paying the price for misdeeds either in the present life or in a previous life is not the whole picture. It can be true in some cases or partially true but it could just as well be an advanced soul who has taken on these problems in order to gain mastery. In that sense, the person (their soul) drew in the problems for this reason. They could be very near to enlightenment. As well as this there are certain things that happen to us in our life to enable us to master Love which are to do with identifying what is <u>not</u> Love, saying 'not that' 'not that' to fear, jealousy, sentimentality, possessiveness, pity, etc. When you see, actually perceive, the falseness of something, then you know Love. Some of the aforementioned emotions are often mistaken for love but they have nothing whatsoever to do with it. Again, it takes absolute honesty and a certain state of awareness. So it is very important not to judge the life of another based on outer circumstances. The soul will always have its reasons for circumstances and these are not always obvious.

Reincarnation or rebirth is no more than dynamic life. We see it in nature with the seasons. Each year the tree bears new leaves but it is the same tree. The bulb produces new flowers each year but it is the same bulb, each time appearing to undergo a rebirth. The eternal part of us remains but takes on a new body for another season. In this way we eventually learn all our lessons, the ultimate one being how to Love unconditionally on every level of life. To master Love. To be a master or have mastery means to have control over or be a skilled practitioner in something, in this case the self, including the mind, senses and physical body. It doesn't mean to have control over someone else, which is another way the word master is used, but that is never the way it is used in a spiritual context. It's rather like the great artists of the past which we refer to as the masters. Spiritual mastery of the self is both an art and a science and this is achieved through the long evolutionary process, the ultimate aim of which is to master Love. When we do so we radiate coherent light waves which enables us to be centred, or to be in the centre of the spiral. Another way of saying it is we become One. This process is also known as Self- (meaning the Higher Self or Soul) realisation, We become One with the Source, or in physics terminology, we experience non-locality as a conscious state.

CHAPTER 22

THE HEART AND QUANTUM CONCEPTS

> *"The brain has its own sense of humor that tends to be more controlling, cynical and sarcastic, but when our heart laughs, our eyes often flood with tears from the recognition of the divine wonderfulness of the futility of our attempt to approach life's challenges exclusively on a rational basis."*
>
> Paul Pearsall, Medical Doctor, Speaker and Author

The heart seems to be the focal point in the human body for experiencing love. As Sara Paddison,. Vice-President of the Institute of HeartMath says "Whatever your religion or cherished beliefs, the heart is the access point in the human system for experiencing God." As described earlier, the heart seems to be the dominant organ in terms of electromagnetic, or light, radiation. In his book "*The Heart's Code*", Dr. Paul Pearsall notes research where he was able to demonstrate that when someone is exposed to love and compassion, their heart radiates light more coherently than when exposed to violence (correlating to Dan Winter's experiments with ECGs). He was also able to demonstrate that cells, all cells, have memory, and this came from his work with heart transplant patients. This is consistent with the theory of the Morphic Resonance of Light. If you remember it was explained that memory exists in the morphic fields operating at faster speeds of light. The ultimate speed being the description of relativity where past, present and future co-exist, which would contain all memory.

His discovery for this came from the recipients of heart transplants who took on the characteristics and memories of the donor. The most startling case was a young girl who received a donated heart from a young girl who had been murdered. After recovery from the operation the young recipient had an experience which terrified her as she told her mother that she knew who the murderer was. With her descriptions the police were able to catch the murderer and convict him. What's happening here is that the memory field is still in existence because the heart is still part of the original morphic resonance field. However, eventually the memories fade as the heart becomes part of the resonance of the new body.

Also in Paul Pearsall's book he describes experiments which have proven the healing effect of love in physiological terms. He writes of experiments conducted by D.C. McClelland and J.B. Jemmott into immunological effects of a caring response or compassion. [1] One experiment consisted of showing a documentary of the work of Mother Theresa to a group of students and another group were shown a film about Nazi atrocities in the Second World War. The group who were shown the film about Mother Theresa showed a significant enhancement of their immuno-deficiency compared to the other group. [2] Paul Pearsall quotes "Watching a person with a coherent, open heart, what McClelland called 'affiliative trust' seemed to make the students' hearts more coherent, open and more able to tap into their own heart's affiliative trust code, and these changes were accompanied by a lasting enhancement of their immunity." [3] So we see here the effects of love on the coherence of heart rhythms and in turn, the effect this creates on the immune system. This is confirmation of Roger Coghill's work referred to earlier, where he found that what he called the endogenous electric field emitted by the brain and heart were crucial to the vitality of white blood cells.

Taking the above experiment one step further, another study documented in the same book was an experiment conducted by the U.S. Army Intelligence and Security Command (INSCOM) in 1993 where they took white blood cell samples from the inside of the cheek of a volunteer. These were placed in a test tube and hooked up to a recording polygraph machine. The volunteer donor of the cells was separated in a different room in a different part of the building from his donated cells where he was shown a violent film. At the same time this was happening the recording polygraph showed wild fluctuations from the cheek cells, consistent with what would happen if the polygraph was attached to a person who was undergoing emotional turmoil. The experiment was repeated using greater separation of distance and time between donor and cells but they consistently showed the same results. In other words, the cells behaved as if they were the person, which means they not only had consciousness but separation of time and distance did not affect the non-local connection. [4] This is a perfect example of the nonlocality of consciousness.

Even older studies exist which show this same non-local consciousness. These were conducted in the 1960s by Cleve Backster [5] and reported in the book '*The Secret Life of Plants*' by Peter Tompkins and Christopher Bird, published in 1973 by Harper and Row and referred to in earlier chapters. The studies reported in the book showed that different body cells exhibit the same non-local connection to their donor and that space and time of separation do not alter this connection. But even more startling studies showed that plants also have consciousness and that they react, demonstrated by polygraph recorders, to the thoughts and intentions of the people around them, particularly when the intention is towards the plant itself, such as the thought of harming the plant, for example. This is showing not only non-locality in action, but that ***everything*** has consciousness and it is all One consciousness. This is further evidence of the Morphic Resonance of Light, where everything is made of light, that the light has consciousness and it is non-local. In other words, omnipresent.

Another example of an experiment showing this non-local consciousness is a more recent study which also further shows the effects of love and compassion to create coherence, hence health and healing. This experiment is also documented in Paul Pearsall's book and consists of a prayer study conducted by Dr. Randolph Byrd in San Francisco. The study consisted of heart surgery patients who were prayed for by various prayer groups around the world. They were found to have better recovery rates than the control patients who weren't prayed for by these groups.[6] In his book he reports more than 70 such studies which have shown the same non-local effects as the cheek cell experiment.

If this proves anything, it's that non-locality is a factor of human consciousness, but also that there is alignment with a healing force. We've already seen above, that coherent heart rhythms more easily enable a centreing within the spiral, or a coherent nesting of waves, and that this produces an enhanced immune function and it is Love which is producing the coherent heart rhythms. So, it follows that those praying are aligning with Love (in whatever form or name they may choose) which, through non-locality, creates coherent heart rhythms in the patient, enhancing health. But it has been shown through double-blind, scientifically controlled experiments run at Harvard University that prayer is effective even when the people praying and the people being prayed for have never met and are located in different parts of the world. This really shows that there is no such thing as separation, that we are all connected to everything else, which really means we are all Source. That's what non-locality means in its ultimate sense. The non-locality of lightwaves or conciousness means we are all One, and by 'we' I mean the human, animal, plant and mineral kingdoms and all inanimate matter– seen

Light in other words, as well as all the unseen Light. Saying that everything is made of the lightwaves of consciousness means that everything, including tables, chairs, boats and planes are all made of consciousness. It's all consciousness, which is the same as saying It's All Light. The only difference between animate and inanimate matter is the creative will, which can express to a greater extent the more evolved the consciousness is. Creative will expresses through what was earlier called the soul or the hologram and in physical existence, humans are the highest form embodying creative will though it can be expressed to an extent from the lower kingdoms, particularly animal, but even plants and minerals have anima (life or soul force) though they obey the consciousness of stronger forces. As in the double slit experiment referred to earlier, the photon either creates two lines or a wave interference pattern thus showing that a) it has consciousness because it made a decision and b) that it is creative because it creates either the two lines or the waves, consistent with the requirements of the laws of the third dimension. So creativity is inherent in consciousness. Creative will is simply the logical and natural outcome of the evolutionary process and is a function of intelligent consciousness. As the consciousness evolves ever closer to its ultimate Source, which is Creation itself, then naturally there would be a greater expression of creativity.

Morphic Resonance of Light is not only the simplest explanation of the forces in physics but provides an explanation of the above experiments. It is really showing that there is no such thing as separation. Everything is non-local resonating light. It also provides an explanation of biological mysteries such as the structure of form, in chemistry it provides an explanation of chemical reactions because chemicals are made of sub-atomic particles which take their place in the morphic resonance patterns which are unique for each chemical. This is the same as saying that each chemical, like everything else, has its own unique sound, (or frequency) as proved by Jacques Benveniste and others. And it provides an explanation for spiritual phenomena such as near-death experiences, clairvoyance, remote viewing, de and rematerialisation and the healing and prayer experiments mentioned above. It is the simplest explanation for describing all of life from the sub-atomic particles to the super-clusters of galaxies and all the unseen dimensions.

It is all dynamic light resonating to harmonic frequencies. The light coming in, producing matter, and going out, producing radiation.

CHAPTER 23

AURAS AND RADIATION

> *"Scientists G. Schwartz, PhD and L. Russek, PhD, who are pioneers in the field of energy cardiology by integrating modern science with the wisdom of the ancient systems of healing point out that medical science has been afraid to take the logic of the physics of ECGs and MCGs to their inevitable conclusion. These fields are measured outside the body which means the biophysical energy reaches the skin, whereupon they argue it doesn't just simply stop. It continues to radiate out into space. Simple physics tells us this. So as humans we are radiating energy continuously."*
>
> Paul Pearsall, Medical Doctor, Speaker and Author (extract from 'The Heart's Code')

When a person is radiating coherent light waves it produces a healthy aura, or biophoton field or endogenous electric field, whichever terminology one wishes to use. The aura is composed of the energy fields, really electrical and magnetic fields, of each individual hologram or soul, of which the physical body is a precipitation. As discussed above in the chapter on Electromagnetism, we are electromagnetic beings, every sub-atomic particle is charged with electrical energy constantly coming in and radiating out, so we are radiating electricity, and through the spin of electricity in our vortices, magnetism. Present day aura photography can only capture a certain portion of the aura but as technology advances it will become possible to capture a more accurate picture of the harmonic levels of the aura. The higher harmonics are simply beyond present technology to measure.

The aura is composed of many colours which are, in fact, sound at different frequencies. If you remember what was said earlier about when one octave from lower F to upper F is played 40 octaves higher it produces the light of our colour spectrum. Each particle is, remember, a particle of dynamic light. Light, as we know, contains within it all colours, each resonating to a different frequency. In the visible spectrum, red has the lowest frequency (longest wavelength) and violet has the highest frequency (shortest wavelength). So the various resonating frequencies will release a different colour. Remembering that resonating means resounding, this is why some people say they can hear colour and why some people can see sound. The aura is dynamic and interactive, constantly changing as energy comes in and goes out of the energy vortices in the body, not only the sub-atomic vortices of each atom but also the 'chakras'. The chakras are spinning vortices of light located up and down the body from the frequencies of red (base of spine) to violet (top of head). These are the fundamental frequencies although other frequencies come and go as the whole system is dynamic. Chakra is a Sanskrit word for 'wheel' as that is what they look like, spinning wheels. Each chakra is a spinning vortex and they have smaller vortices within them, like wheels within wheels. The seven major chakras in the body are each linked to a major endocrine gland.

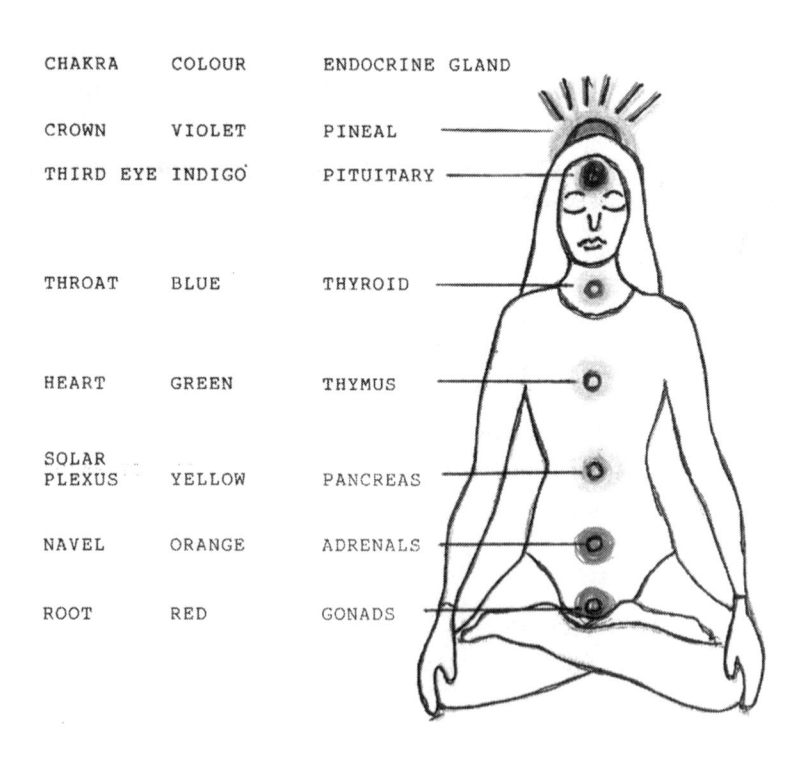

CHAKRA	COLOUR	ENDOCRINE GLAND
CROWN	VIOLET	PINEAL
THIRD EYE	INDIGO	PITUITARY
THROAT	BLUE	THYROID
HEART	GREEN	THYMUS
SOLAR PLEXUS	YELLOW	PANCREAS
NAVEL	ORANGE	ADRENALS
ROOT	RED	GONADS

FIGURE 11 – THE CHAKRA SYSTEM OF THE HUMAN BODY

The endocrine glands receive their energy, their vitality, from the interplay of energy within the chakra. That is why many healers work on the chakras. If the chakra is functioning well the linked organs will function well. There are many smaller chakras throughout the body also, for example on the hands and the knees, but it is the dynamism of this whole process which makes up the aura and the whole chakra concept is consistent with the vortex and fractal principles.

The halo, as depicted by artists around saintly figures, is an artistic impression of the light or aura seen around people. This radiatory light emanates from all beings to varying degrees dependent on a person's point of evolution or their ability to consciously 'tune into' and realise higher realms and thereby radiate the energy. It is the radiation of this energy which quantum physics also tells us that all sub-atomic particles (vortices) radiate. We know that electrons release photons which are then absorbed by other particles. Everything around us is in constant exchange so all life is dynamic and interconnected.

The radiation of auras is energy that is radiated as the spiralling lightwave is released from the atoms. When consciousness is highly tuned, in other words tuned to the higher octaves of the harmony, then obviously the radiation released is of the higher octaves also. That is why healing can occur simply by being close to a saintly person. So this release of energy is radiation, that is what radiation means. The radiation is spiralling lightwaves. This is seen as subtle light for those who have clairvoyance. This radiatory subtle light is what makes up the aura. The aura, as evidenced by those who can see, can vary in the extent of its radiation depending on the evolutionary level of the soul. The aura's colours can change moment to moment, according to one's thoughts and feelings because the thoughts are attuning to different consciousness levels. The colours themselves have meaning and different qualities. Of course, in science we learn that a prism breaks up white light into all the colours of the rainbow, each colour having a different frequency (or wavelength). It is obvious then, that each colour will have a different effect on consciousness due to it having an effect on subatomic particles (the vortex), hence on atoms which has an effect on cells. This is the basic premise of colour therapy. But there are many different octaves of colour and what we see in the physical spectrum is but a piece of the entire range of colours. Although there are some highly effective colour therapists, the greater understanding of colour is yet to come.

The topic of radiation has been discussed in relation to spiritual light, but the word radiation is used by us in a more ominous way also, that of the splitting of the atom, atomic energy or nuclear energy. The word radiation is appropriate in both cases and it is the same radiation but the damaging type is due to mankind's tampering with the atom without the necessary understanding or evolvement, as is true with many applications of science today. The radiation that is released from the splitting of the atom is the captured spiralling lightwaves which makes up each vortex, which shows us the potential of energy contained within each vortex.

Radiation is really a factor throughout all of life, throughout all the kingdoms in nature and throughout all the dimensions from the physical to the highest spiritual. There is a radiatory edge between all the dimensions, one blends into the other and the difference between the dimensions is in the degree of radiation. This can be easily understood when looking at the electromagnetic spectrum. It is my theory that what scientists have done is release the inherent Life energy (the stored spiral energy within each vortex/subatomic particle) through artificial means, what we know as nuclear energy, without the necessary evolvement or state of consciousness. They have forced what should be a natural evolvement. It is my understanding that the highest spiritual being would be able to withstand any amount of radiation because their consciousness is tuned such that they can release it. They are dynamically tuned to the faster speeds of the spirals/light waves and so would be capable of releasing vast amounts of light as a result. We on the physical plane, cannot withstand such an intake of energy because we are tuned to a lower speed of release of spirals/light waves. Our sub-atomic particles, hence cells, cannot release, or radiate, the necessary spirals/waves fast enough. This usually becomes cancer. It is really to do with tuning; where we are tuned to on the electromagnetic spectrum.

CHAPTER 24

THE CAUSE OF CANCER

> *"There is always an optimal value beyond which anything is toxic, no matter what: oxygen, sleep, psychotherapy, philosophy. Biological variables always need equilibrium."*
>
> Gregory Bateson, Anthropologist, Social Scientist and Cyberneticist (1904-1980)

It seems that when something interferes with the natural harmony between spin of particles and their frequency this creates incoherent light emissions and ill health results. Anything which causes particles to maintain an excess of light relative to its frequency will create incoherent light and can lead to overgrowth of cells. There are several factors which can cause a particle to maintain an excess of light:-

Frequency

Inherent in this whole theory is that not only do particles vibrate, which we call frequency, but they also rotate, spin on their axis (thereby creating a vortex) and this effect is created by the morphic resonance. We know that sound waves are a series of vibrating waves. The rate of vibration is controlled by and attuned to the particular resonance. The lower octaves i.e. the physical dimension, not only spins slower but also vibrates slower. A sudden increase in vibration would cook a particle. This is the principle of temperature increase. Increase in vibration causes increase in temperature. If our subatomic particles and on a larger scale, cells, were suddenly subjected to a higher vibratory frequency it would cause overload, or burnout, in other words, we cook! We know this, of course, simply from experience but it was visibly demonstrated by Hans Jenny in his cymatic experiments. He showed that the resonating pattern is damaged by heat. So in other words, the morphic resonance of the third dimension is broken down by frequency (heat). Which adds proof to the statement made above that spin and frequency are linked in each dimension. If the frequency/heat increases without the necessary increase in spin then damage results. This relationship is crucial to health and it is a very fine balance, most easily realised in the fact that despite the wide variety of environments on the Earth where humans live, the body temperature always remains close to 37.5C and this is necessary for health to be maintained. Any slight deviation and ill health results. In the short term the damage would be entropy or disintegration of cells which is the result of increased temperature which is what happens in cooking and also in fever. A fever is the body's way of killing invading bacteria and viruses. Most people suppress a fever at the first sign, not realising that this allows the invading pathogens to survive, thereby prolonging the illness. It has to be observed carefully, however, and not be allowed to rise too high but in most cases this does not happen. The body has wonderful built-in mechanisms to maintain homeostasis.

So although there are thermal effects from faster frequencies there can also be thermal effects from slower frequencies. The best example to demonstrate this is microwave ovens. Microwaves have a slower frequency than our visible spectrum but can be used for cooking due to power density. The way microwave ovens work is that alternating electromagnetic waves are generated inside the oven by an electron tube called a magnetron. These waves bounce back and forth between the metal sides of the oven constantly passing through the food. The waves don't penetrate metal but they do penetrate glass, plastic, ceramic and the food itself. The water molecules in the food constantly orientate themselves to the alternating em waves, causing them to rotate. This is because water molecules are dipolar, meaning they have a positive and negative pole. Also, ionic compounds in food (which are dissolved salts) are caused to collide with other molecules due to the alternating em waves and it's the friction due to the rotation and collision of molecules which generates heat. Another example is mobile 'phones which operate on the radio/microwave end of the electromagnetic spectrum. Again, they can cause thermal effects when placed directly on the skin because of the intense concentration of waves (remembering that living tissue is made mostly of water)..

These are examples of faster and slower frequencies creating entropy or disintegration. Because the human body and other bodies are so finely balanced in terms of temperature, this tells us that inner temperature is crucial to health. Even minute temperature differences could cause biological effects, at levels which are perhaps unmeasurable by present standards. This principle of subtle temperature in relation to health was well understood by Victor Schauberger who said -:

"In Nature all life is a question of the minutest, but extremely precisely graduated differences in the particular thermal motion within every single body, which continually changes in rhythm with the processes of pulsation.

This unique law, which manifests itself throughout Nature's vastness and unity and expresses itself in every creature and organism, is the Law of Ceaseless Cycles that in every organism is linked to a certain timespan and a particular tempo.

The slightest disturbance of this harmony can lead to the most disastrous consequences for the major life forms.

In order to preserve this state of equilibrium, it is vital that the characteristic **inner** temperature of each of the millions of microorganisms contained in the macroorganisms be maintained."

But faster or slower frequencies, if not intense enough to create heating (at least that can be detected), but nevertheless sustained, can cause other biological effects including overgrowth of cells leading to cancer. The reason for this is due to the lack of synchronisation between spin and frequency of particles. As you extend further out from the visible spectrum which the human body is attuned to, the frequency or spin of our particles loses more and more synchronisation. This is why ultraviolet and infrared radiation are less harmful than X-rays and microwaves/radiowaves, because, if you refer again to Figure 2, you will that they sit in a similar position on opposite sides of the electromagnetic spectrum.

Radiowaves which are usually harmless to us, can create biological changes with increasing intensity so it can be harmful to live next to radio transmitter masts, similarly with TV masts or mobile phone masts. The problem arising from mobile phone masts is that they are springing up all over the place, in town squares, on buildings, near homes and schools. Certain organisations, particularly those with a vested interest, are insisting that they cause no harm. This is largely based on the studies showing that there are little or no thermal effects from masts. The mobile phone themselves, however, do cause heating or thermal effects and the intensity of the radiation is much greater because it is being received directly into the head. But many scientists have found evidence of biological effects from non-thermal radiation (non-thermal as far as can be detected) and evidence of health problems in people who live near military communication bases, which use radar and microwave radiation, mobile phone masts, radio and TV masts. [1] While we enjoy the convenience of all these technologies, prudence would suggest keeping transmitters as far away from living organisms as possible.

The reason why these frequencies create health problems is due to the disruption to the synchronisation of spin and frequency of particles. This synchronisation results from the dynamism of the vortex, the taking in and the releasing of the spiralling lightwaves. If the frequency is increased, the spin would be relatively slower. This means the incoming spiralling lightwaves of the particle would be withheld longer than they should be. Because the spin of the particle is slow relative to the frequency, this could create distortion of the morphic field because it would have a type of heating effect (refer to Figure 8 where we saw how applied heat to the cymatic pattern distorted its shape), but it could also mean that the radiatory light of the particle would not be released as quickly as it should be. This means the light would be withheld in the particle and this would create incoherence. This withholding of light would create overgrowth of the cell because that is what light does. It creates Life. It would simply be growing more Life, which is what cancer is. If, on the other hand, the frequency is slower than normal, this means the particle will be spinning faster than the frequency. This could still create withholding of light because its faster spin means it would be absorbing more lightwaves but wouldn't be able to release them quickly enough due to the rate of frequency so there would still be a lack of synchronisation and incoherent light emissions. Once again the withholding of the light due to it not being able to release it quickly enough would cause overgrowth of cells, the creation of new life, or cancer.

Another problem with mobile telephone masts is to do with the way the signal is transmitted. It can often be a pulsed frequency. This is especially true of the newer generation of mobile 'phones. Normally when a cell is subjected to a sudden burst of energy, it will absorb this energy then after a short while homeostasis will kick in and the cell will seek to readjust itself to its normal state. So it is possible that any effects will be short-lived. This is true also for the sub-atomic particle. But when the energy comes in frequent bursts as it does with pulsed frequencies and this is sustained, the cell, or sub-atomic particle, is continually stressed but continually trying to maintain homeostasis. It keeps trying to pick itself up but gets knocked down again, and again and again. It's easy to see that this would create stress and, therefore, incoherent light. If on the other hand, the energy is steady and sustained, perhaps the cell or sub-atomic particle adjusts slightly to the new rate of spin and adjusts its frequency accordingly to maintain a new level of homeostasis. But there would be thresholds or limits to this and it would still have to be within our normal range of attunement. Further complications would be thresholds which varied from individual to individual.

Speed of spin

The spiralling lightwaves, being Life energy itself, in a normal interplay dynamically produces Life as we know it. When a greater amount of spiralling lightwaves are entering the vortex (the speed of spin) but the vortex is unable to release the spiralling lightwaves fast enough, because it's vibrating to a lower frequency for any number of reasons, then the spiralling lightwaves, which is Life energy, is held within the vortex unable to be released. As stated above, this Life energy does what Life energy does, it creates Life, it starts to grow. It is Life energy growing more Life because it cannot be released. To support this argument I can cite some amazing discoveries about cancer. Cancer growing in a body becomes an entity in itself, spreading itself and growing wherever it can. This is what Life does. Some cancers have been known to have teeth and hair, particularly ovarian cancers. Quite extraordinary, yet in view of the fact that it is a Life growing this makes sense. A clinic in London is doing important work in this area using Kirlian photographs to capture images of the Life field, sometimes called the bio-magnetic or etheric field of the body. [2] The Kirlian photographs clearly show that light emanations from a cancer patient's fingertip (shown on the right in Figure 12), for example, extend much further than those from a healthy patient, though the strands of light are tangled and incoherent compared to the coherent smaller emanations from the healthy patient. But this all makes sense. The cancer patient has extra Life energy but which is not being expressed in a healthy way, or being suppressed, creating incoherent radiation, whereas the healthy patient does not have the extra suppressed Life energy and their expression of Life energy is harmonious and coherent.

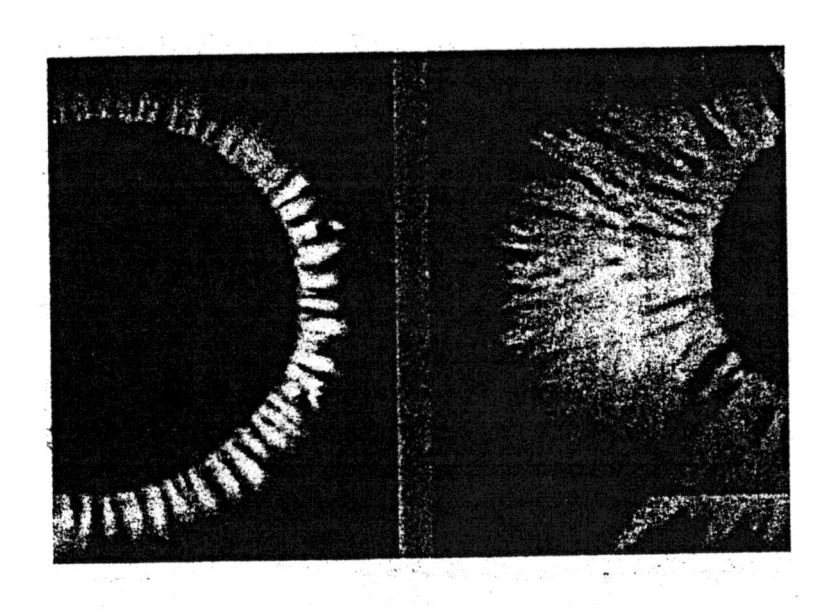

```
Person without            Person with
    cancer                   cancer
```

FIGURE 12 – KIRLIAN PHOTOS OF THE FINGERTIPS OF A PERSON WITHOUT CANCER AND A PERSON WITH CANCER

To further support this argument about suppressed light or incoherent light being the cause of cancer, the work of the Dr. Fritz Albert Popp, beginning in the 1970s, discovered that living tissue emits light (he calls this biophoton energy) as was later verified by Hameroff and others. He also discovered the phenomenon of 'delayed luminescence' and realised it was a type of corrective device for the cell because after a short time the cells radiates more than normal suggesting that the amount of light contained within the cell must be a defined amount, a precise balance. When the cell has received additional light, as in the experiment, it undergoes adjustment then simply ejects the excess. This is the same phenomenon that is happening with the heavy particles (baryons) created in particle accelerators. They initially absorb the excess lightwaves, through acceleration, then when they slow down they radiate the excess light and return to being protons. Baryons, I would suggest, are just protons with extra light gained through speed.

Fritz Popp conducted many experiments on light and was interested to find what carcinogenic (cancer causing) chemicals would do when subjected to light. In every instance he found that when subjected to UV light, the carcinogenic chemicals would absorb the light then radiate it at a changed frequency. The chemical changed the frequency. He discovered that the carcinogens reacted specifically to light frequency of 380 nanometres and was able to predict which chemicals were carcinogenic simply by their reaction to this frequency. He also discovered around this time a phenomenon in biology called 'photo repair' which is the method by which cells repair their damage. If a cell is subjected to high intensity UV light resulting in near destruction of the cell, it can be repaired within a very short period of time simply by being subjected to the same wavelength but at a very weak intensity. This process of cell repair works best when the wavelength is 380 nanometres. What he had discovered was that carcinogens scramble light at 380 nanometres making it incoherent which interferes with the photo repair system of the cell. He believed he had found the cause of cancer and published a paper with his discoveries. [3]

Realising the implications of this finding and with the help of a student, Bernhard Ruth, they built a machine which could count individual photons He experimented with counting the number of photons emitted from people, plants and animals and discovered that there was a steady stream of photons being emitted from all living organisms. (This is further proof of the aura or biophoton field). He used his machine to measure the emissions and discovered humans emit light according to natural biological patterns, occurring in repeating cycles and this is consistent with periodicity or frequency being inherent in all of nature. When he compared the studies of healthy people against people with cancer he found that the cancer patients' light emissions did not follow the biological patterns and that their light was incoherent. (This has implications for people who do shift work as they are not following a normal biological pattern of night and day). He went on to study light emissions from seedlings, some grown in light conditions and some in the dark, such as potatoes, to account for the factor of photosynthesis. He discovered the light emissions from seedlings was the most coherent light he had seen and was astonished to discover that the light intensity from the potatoes (grown in the dark) was even higher than the seedlings grown in light (which meant it was not dependent on photosynthesis). [4] This tells us that -

    a) seedlings have more coherent light than adult plants – this makes sense in that they are full of the potential of the morphic field. Remember Harold Saxton Burr's discovery that the morphic field exists as a template around young organisms and they 'grow into' it.

    b) subtle or unseen light is a reality

Experiments of this nature were also conducted in the 1930's by an engineer who worked for the Kansas City Light and Power Company, T. Galen Hieronymus. He was able to prove that plants can grow using non-visible, or subtle light, which is transmissible down a copper wire. His experiment consisted of placing a metal plate on his roof which was exposed to sunlight and attaching a wire from this to a metal-lined box containing growing seeds in his dark basement. The control group of seeds had no connecting

wire and didn't thrive so well whereas the ones connected by the wire grew into strong healthy plants. This tells us that -
   a) lightwaves and electricity are synonymous, as described earlier. Or electricity is the unseen light which we can harness and transmit by wire
   b) subtle or unseen light is a reality..

The fact that the potatoes had such a high intensity of light is probably due to the nature of the species of plant. Potatoes are a plant which are reactive to visible light. This is why they become green when exposed to light for a while. The green colour is due to a chemical called solanine which is quite poisonous to us. So they have to be kept in the dark to avoid this. But remember, the dark is Light. It is full of the unseen (or subtle) light, so this could account for why potatoes especially would have a high (subtle) light intensity when measured.

Fritz Popp went on to do further experiments on all kinds of foodstuffs and found that, for example, free-range eggs had a more coherent light than battery eggs and, perhaps not surprisingly, found that the healthiest food had a lower and more coherent light emission. This corresponds to the Kirlian photographs of the cancer patient's fingertips. The cancer is suppressed Life energy which radiates incoherently. Because it is suppressed, it is held back, dammed up, so there is more of it but because it's suppressed it radiates incoherently. A useful analogy would be to imagine a waterway where a blockage had occurred due to a build up of matter, say a dead tree, then comes autumn and leaves accumulate around the tree and perhaps some litter floating downstream becomes entangled in it. If you can visualise the water flowing down the stream, where the blockage occurs there would be a build-up of water behind it and whatever water can get past will be in trickles and gushes rather than the smooth flowing which occurred before the blockage.

So suppressed Life energy creates incoherent light, which can lead to cancer, if uncorrected. We need to ask then what creates suppressed Life energy or incoherent light. The incoherent light means that the Life energy is not being expressed correctly. We know that carcinogenic compounds create incoherent light. Spin and frequency out of synchronisation also create incoherent light. As to why this happens we need to look further as it falls into several categories:-

a) With regard to genes. There are some in the medical profession today who regard genetic inheritance as the ultimate cause of cancer. The truth is unlikely to be so clear cut, however, since there are many examples of people who develop cancer who have no history of it in their family, and conversely, people who live a long, healthy life free of cancer when there is a history of it in their family. Also the well-known phenomenon of people from a relatively cancer-free country migrating to a developed country and adopting a typical Western diet, who then succumb to the typical cancer rates of people in their adopted country. There is clearly something more going on than genetic influence here.

b) In regard to exposure to pollution. Electro-magnetic pollution from both ends of the electromagnetic spectrum can cause spin and frequency of particles to be out of synchronisation, creating incoherent light. As for chemical pollution, as was shown in the water crystals experiments of Dr. Emoto, pollution causes injury to water crystals which would affect the correct transference of frequencies in the body. This would create lack of synchronisation between spin and frequency causing the lightwaves to be incoherent. Or, as with Fritz Popp's discovery, if a carcinogenic chemical is encountered it will be changing the frequency of 380 nanometres which allows a cell to repair itself. As far as food is concerned, all food starts off from the same sub-atomic particles so it all has the same quota of life energy to begin with. Along the way, when food is processed; this can include the addition of chemicals, (of which there are thousands used by the food industry), drying, frying, microwaving, irradiation etc. this interferes with the synchronisation between spin and frequency of the particles, creating incoherent light.

c) In regard to exposure to nuclear radiation, as explained above, the forced release of the Life energy inherent in every subatomic particle of every atom, because it radiates away as nuclear radiation, will be absorbed by any particle within its reach, for example in the human body. So the captured excess of spiralling lightwaves, unless it can be released by the particle in the human body, will grow new life. So we get cancer from radiation. This is the same principle as already described. The extra light causes the spin and frequency of the particle to be out of synchronisation, creating incoherent light. We are all exposed to radiation all the time, however, from cosmic radiation and radon from the Earth, but this is unavoidable and usually within limits the body can tolerate. There are places, however, where radon is released in high concentrations from the Earth's mantle and steps should be taken to avoid too high a concentration. The problem really comes from there not being a safe level of radiation due to the factors explained, so it would be wise to be prudent and avoid unnecessary radiation

d) In regard to personality types, the cause is still the same - an excess of lightwaves which cannot be released fast enough, so causing new Life to grow. The personality types fall into two main categories -

   1. Selfish - A person with selfish tendencies just by their very nature tends to withhold, they take in more than they give out. We have already discovered how everything is made of the same stuff – lightwaves, visible or invisible. It does not matter whether the

person hoards money, material goods, help, goodwill because they often all go together in the selfish individual – the principle is the same. It is the withholding of energy and lack of release of energy which will cause the particles to withhold the light instead of radiating it at the correct rate, therefore creating lack of synchronisation of spin and frequency causing the overgrowth of Life.

2. Oppressed - Manifesting as shyness, inhibition or resentment. It is said in esoteric teachings that gifts from the Creator which are unused will become disease. This is because gifts, whether they be poetry, music, art, writing, singing, the ability to heal, the ability to be empathic, or any number of other things, means that the person is in tune or in touch with a higher aspect of reality. But these gifts are not limited to the few. The potential is inherent in everyone because we are all part of one consciousness. We all have specific talents and abilities, some more pronounced than others but when they seem to be lacking, the clue is being open to inspiration. When we are open to inspiration we can absorb faster spiralling lightwaves of consciousness, this is what inspiration is. If gifts are suppressed or not able to be expressed fully, the Life energy is held back in the subatomic particle and does what Life energy does, it grows!

This can also explain what is happening with the battery hens mentioned earlier, whose eggs radiate incoherent light as opposed to coherent light radiated from free range eggs. A convincing argument for choosing free-range eggs and chickens I reckon, although from an animal welfare point of view, this is needed evidence to show how miserable and unhealthy intensively reared animals are. Any organism which is denied the ability to fully express its nature will feel oppressed. This oppression causes Life energy to be suppressed, thereby creating incoherent emission of light because the spin and frequency of particles will be out of synchronisation This oppression also goes some way to explaining why 'nice' people get cancer, a well-known phenomenon by the medical profession. They are 'nice' because they are constantly bending to others expectations or will, and so afraid to say 'no' or to just be themselves, to express their own nature. They are oppressed and therefore feel suppressed. This is not to be confused with the individual who bends to others 'needs' where a life of service to humanity or the world is the fast track to evolution. This is performed from a duty of Love and that person will therefore be radiating coherent light. This person can also be a 'nice' person. But I am talking about the individual who feels coerced in some way into carrying out someone else's will. This can be the will of either a person or an authoritative body, such as a religion for example. The pressure to be seen to be 'good' and the repercussions of not being 'good enough' can drive an individual to perform good works but with fear or resentment in the heart rather than joy. The good works can often be done to seek approval, recognition, acceptance or respectability and the coercion so subtle and so in-built, through early conditioning, that it goes unrecognised. Resentment is defined as holding back anger, anger which is held in rather than expressed, in other words suppressed energy. It can range from niggling inconvenience to moderate annoyance to extreme anger, but it's all still resentment. As explained, suppression creates incoherent light. This is why long-held resentment can cause cancer. This is a well-known fact in natural healing and mind-body medicine circles.

One such study conducted nearly fifty years ago was the relationship between cancer and withheld emotions where a group of smokers were compared, some of whom were suffering lung cancer and some were suffering other diseases. It was found that the cancer patients had poorer outlets for emotional release and the study concluded that "the more repressed a person was, the fewer cigarettes were needed to cause cancer" from *'Love, Medicine and Miracles'* by Bernie Siegel M.D.,p80 So suppression by whichever cause, whether it's selfishness or whether it's oppression - resulting in resentment, shyness or inhibition, (the root cause of all of these is fear), prevents a person from **fully** expressing the gift they have. Or another way to say it, they are prevented from fully expressing who they are. Wrong education is really at the heart of this problem. So often we are told, in so many ways as we grow up, to hide our light under a bushel because others may be threatened, to not be all we can be because we should know our place and not get above our station! Thankfully, with the growth of the self-development and awareness movement in recent years due to this crucial point in our planet's evolution this situation is beginning to be corrected. We are learning now to shine our light, as we are meant to do. This was so eloquently stated by Nelson Mandela in his inauguration speech in 1994.

"Our deepest fear is not that we are inadequate.
Our deepest fear is that we are powerful beyond measure.
It is our light, not our darkness, that most frightens us.
We ask ourselves, "who am I to be brilliant, gorgeous, talented and fabulous?
Actually, who are you not to be?
You are a child of God.
Your playing small doesn't serve the world.
There is nothing enlightened about shrinking so that people won't feel insecure around you.
We were born to make manifest the glory of God that is within us.
It is not just in some of us; it's in everyone

And as we let our light shine, we unconsciously give other people permission to do the same.
As we are liberated from our own fears, our presence automatically liberates others."

It is, therefore, vitally important to let our light shine. Anything which interferes with the speed of spin of particles and their corresponding frequency will create incoherent light. When we let our light shine, it not only keeps us healthy but helps everyone and everything which comes near us due to the radiation of coherent light waves. The resonance of these coherent light waves helps their own centreing process along because of resonance. Resonance means responding to vibrations of a particular frequency by echoing or resounding. So a person will echo or resound to the coherent resonance. In other words, they will echo the coherent light waves themselves and this is why "our presence automatically liberates others".

So many things in our lives touch us though we may not be aware of it at the time and there are countless references in spiritual and religious texts as well as the esoteric philosophy found in the Ageless Wisdom teachings which refer to our true nature, our essence, as being that of Light. So while there are many religious and spiritual paths all leading back to the same ultimate Source, and all equally valid, the only caveat would be that it would have to be a path of Love, by whichever name or method one adopts, it would nevertheless have to be Love. Because Love, as we have seen, takes the form of lightwaves and we have examined the potential of lightwaves The Ageless Wisdom teachings are an ancient body of teachings of unknown origin which have been given to the world through the ages by various teachers; in the recent era from such people as Helena Blavatsky in the late 1800s, Alice A. Bailey in the 1930s and 40s, Swami Muktananda, Swami Yogananda and the Agni Yoga teachings and many, many others. What they, and the other religious and spiritual texts reiterate is that, **we are known by our Light.** The radiation of our Light is what is seen by higher dimensional intelligences.

Fritz Popp did valuable work on light emissions but like so many before him, was shunned by the scientific community for daring to suggest that light is emitted from living organisms, yet he had the proof and if you remember earlier it was explained that atoms absorb and emit light, so why is it so difficult to believe, considering we are made from atoms. But we have moved beyond belief here, he has the actual proof. The good news is that over time, it took about 25 years, he has managed to gather together a group of open-minded scientists who also began to consider that resonance and frequency may be the body's information network. Together they formed the International Institute of Biophysics in Dusseldorf, Germany. Fritz Popp had already discovered where the light was coming from, or at least where a great proportion of it was coming from. In studies with DNA he was able to unwind the double helix using a chemical called ethidium bromide. As the double helix unwound it released light, the more chemical used the greater the light (due to greater unwinding), the less chemical used the less light was released [5] DNA itself is a spiral of light and he found it contained a range of frequencies. Not surprisingly when you consider DNA's role. He called it the master tuning fork of the body. This analogy is very consistent with morphic resonance, if you remember earlier where it was explained that different frequencies actually alter the structure of form. But in the Institute they also found that not only is light emitted from living organisms but that it is communicated to other organisms by exchange and absorption. [6] This is why healing works. The healer is the transmitter through which coherent light can be sent to the patient because the healer is attuning to coherent light. It can also be sent at a distance, any distance, because of the factor of non-locality and this distant healing is a commonly-practiced method in the healing arts. So through understanding quantum physics we can understand how and why healing works.

Other scientists have come up with similar findings on resonance and frequency. The Italian physicist, Renato Nobili, Universita degli Studi di Padova, showed wave patterns in the fluid of cells which correspond to wave patterns in electroencephalogram (EEC) readings. [7] Herbert Frohlich of the University of Liverpool, England, introduced the idea that collective vibration was the carrier of instructions in the body for proteins, amino acids etc. [8] He showed that when energy of the right intensity is reached, this creates coherence amongst molecules and that they take on the quality of non-locality. [9] This is because coherence is communication and information accumulation: Wholeness in other words.

The morphic resonance fields store and contain all information because they are frequency waves. Sound, remember, is waves at particular frequencies and waves are the ultimate carrier of information. Since the spiral shape is the pattern of the universe, waves would have to embed and be coherent in the spiral shape for non-locality to happen. Since the golden ratio spiral seems to be the one where waves can embed coherently this would make it the perfect container for all information. Because it is perfect coherence it means that it is the perfect medium for non-locality.

Non-locality means everywhere at once; there is no such thing as separation. Non-locality is inherent in the morphic fields of consciousness of which we are all a part, the dark matter, the spiralling waves of consciousness, existing in harmonic multiples on and on into infinity. So it is ultimately one gigantic field of consciousness of coherent lightwaves which are everywhere at the same time. A great cosmic mind, as described by Max Planck or a giant thought, as described by James Jeans. For every individual who becomes enlightened, in other words, radiating coherent light waves, it helps to increase the critical mass of the centreing process, thereby enabling more quickly the enlightenment of others.

At the same time as the universe is expanding, gravity is drawing all planets and solar systems together in clusters, so there is a centreing going on at the same time as Life is continually being created. The ultimate outcome of the centreing could be the 'Big Crunch' which is one of the possible outcomes of theoretical physics. This means that ultimately all matter will compress itself together and be destroyed in the process. This, however, would not be the end of everything. It would be the end of a complete breath of Source; the end of another turn of the spiral which will be followed by another in and out-breath on a higher turn of the spiral. The evolutionary process of creation itself can be explained by the spiral shape. Each turn of the spiral being an out breath or devolution – the descent into matter, then on the return an in breath or evolution – the return of matter to spirit. At the end of each complete revolution comes the big crunch, beginning again on a higher turn of the spiral, closer to the centre, closer and closer to the Source. Perhaps there is a new Big Bang, a new OM at each turn of the spiral.

CHAPTER 25

MUSIC OF THE SPHERES

"Music does not touch merely the mind and the senses; it engages that ancient and primal presence we call soul".
John O'Donohue, Poet, Philosopher and Author (1954-2008)

There are many clues left behind in ancient scriptures to tell us how the world was created and they all refer to sound as being the instrument. The Bible has already been referred to; "in the beginning was the Word", and in Vedic philosophy, the most ancient scriptures known, 'The Bhagavad Gita' literally translates as 'the song of God'. The Australian Aborigines believe that the world was sung into being. All of us are held together by sound, indeed everything we see around us is held together by sound. This can be a difficult concept to grasp since matter seems so solid and static with changes only occurring due to time but it becomes easier to understand when using the metaphor of music since timing and change are central to the very essence of music. All the faster dimensions which we cannot see physically are also held together by sound. All of it is resonating and all of it together makes up the harmony of the spheres. As in physical dimensional music, harmonics are made up of different octaves. Each ascending octave being an ascending pitch or tone or, in other words, a shorter wavelength. This analogy follows in the description of ascending dimensions of light waves or consciousness. Each dimension having shorter wavelengths and increased speed of spin of particles, creating slower time. Each dimension being an octave or harmonic multiple of the speed of light, the particles held together by sound but in a higher octave or harmonic. We can't normally see or hear it because it is moving too fast. It is beyond our relative consciousness, unless, as described previously, we can 'raise' or tune-in our consciousness. The 'strings' of string theory is a useful analogy to use here since the spiralling lightwaves being attuned to the morphic resonance can be likened to strings of an instrument being tuned to the correct sound.

Many experiments have been carried out over the years which have shown that sound or music can create either healing or unease depending on whether it creates resonance or dissonance. Some of the earliest experiments were documented in the 1973 bestseller '*The Secret Life of Plants*' [1] which showed that plants either responded well to certain types of music or shrivelled up and withered to other types of music. The music they seemed to prefer were the classical pieces. This is also discussed in the book '*The Mozart Effect*' [2] published in 1997 by Hodder and Stoughton,in which the author Don Campbell presents his findings that music is not just entertainment, it is medicine for the body and soul. Many clinics and hospitals around the world are now using music as a therapy, with good results. This is because music reaches our very core. We seem to know at some level that this is what we are in essence. As the late John O'Donohue says "to the human ear, music echoes the deepest grandeur and the most sublime intimacy of the soul".

It has already been discussed how mathematics correlate to harmonic octaves and musical intervals. This did not go unnoticed by the great thinkers of the past. The German philosopher Gottfried Wilhelm Leibnitz (1646-1716) said "Music is a secret arithmetical exercise and the person who indulges in it does not realize that he is manipulating numbers". In the chapter on Morphic Resonance it was described how Plato in "*The Timaeus*" explained how all life is composed of measure and number, proportion and mathematical form and that this also applies to musical intervals. Indeed he believed that the creation of what he termed 'the world soul' was in accordance with musical intervals and harmonics. This is a good description of the harmony of the spheres. The Pythagoreans also discovered that the cosmos can be described using musical harmonies which are based on mathematics. As noted earlier, Astronomer and Astrologer, Johannes Kepler also had this understanding and worked out that the spacing of the planets was based on harmonic intervals. It was also described earlier how orbital pathways and the 'shells' of electrons are most likely based on harmonics of the morphic resonance field. For example, the morphic field of the sun would create the orbital paths of the planets which orbit it, each orbital path being a harmonic of the nucleus. In this way we can now see how physics, geometry, mathematics, astronomy, music, nature, love, everything, in fact, comes together and is interconnected, all part of one law, which could perhaps be called 'The Law of Love': Love being the ultimate coherence. Since music and mathematics are descriptive of each other, the golden ratio is therefore descriptive of a sound. Since it has been described as being the model of the universe because it is the fractal that is found throughout nature and all life, it could be the OM itself.

It was discussed earlier how the corpus callosum, the bundle of nerve fibres which connect the left and right sides of the brain, is thicker in women. It has also been found to be thicker and more fully developed in musicians, both male and female. This is further confirmation that music is holistic, sound is holistic, OM is holistic.

When the soul descends into matter and the experience of Earth, the coherence, the harmony can become distorted. The whole aim of our evolution is to return to the harmony of the OM. The method by which this is achieved seems to be Love. Love creates coherence which creates harmony. The word harmony comes from the Greek *harmonica* which means 'fitting together'. Or another way of saying it is, Love creates coherence which creates embedding within the harmonic speeds of light. Love creates coherence within the universe.

    Like Earth music, many notes created by many instruments together make up the orchestra. This brings to mind the word compose which is the putting together of various parts and is often used in a musical sense, a musical composition which is a harmony. This is what we try to do when we compose ourselves, we try to create harmony for ourselves, to regain our composure, which is really another way of saying centreing or becoming harmonious or in resonance to our surroundings. The use of words is really no accident. At some level we know that to be composed is to attain resonance with the greater harmony. We are a note in the great orchestra. This is what happens when we gain composure, the dictionary definition of which is 'tranquil demeanour' or 'calmness'. On Earth we are instruments emanating notes, but as explained, we are dynamic and interconnected with all life. We are therefore not just a physical body. The body is the instrument. Because we are dynamic and interconnected, particles created from light waves, in essence, we are the light waves ultimately emanating from Source. Ultimately, therefore, we are all Source, everything is. We do indeed have the kingdom of heaven within us, as Jesus and others have taught. The idea that we are separate from each other is an illusion of the third dimension. It takes a leap in consciousness to realise this but the groundwork is already covered with the findings of quantum physics, particularly the findings of non-locality and the work on coherent heart rhythms by Dan Winter, the HeartMath Institute and others. Another way of saying it is consciousness has to be 'raised', or quickened/speeded up, to realise this, not believe it, but REALise it, that is… to make it real. A way to realise it is to practice feeling love and compassion. In a world that seems to be going insane it can be difficult to do this but one way is to identify what is <u>not</u>

Love, as noted earlier, or to practice meditation which brings the brain waves into a state of coherence, as noted earlier. There are many ways of achieving meditation and while transcendental meditation was mentioned earlier and this is taught as a method, it should be noted that all meditation is transcendental. The great philosopher, J. Krishnamurti, noted that meditation is not a system or a method as these are mechanical things. Meditation is the opposite of mechanical so it needs to be arrived at by natural means   By that I mean letting go of the ego, especially thought, which carries all its attachments, knowledge, etc. When this is let go of then the state of ego-lessness or One-ness is achieved, what some may call Zen. But since this state is far from what many experience as normal it may take some discipline to achieve and this is where help may need to be sought initially.

There are also many healing therapies available which can open the heart to love. Some of the more effective ones in my experience are spiritual healing and emotional freedom techniques, the latter being particularly effective in clearing out old conditioning patterns, attachments and limiting beliefs, which are all of the ego state, to allow the true nature to shine. If you remember, emotional freedom techniques (EFT) was mentioned in Chapter 5   Holistic Consciousness. It is very effective in physical healings due to the effect that emotions have on the chemical interactions of the glands and the immune system, as proved by the studies of psychoneuroimmunology, but it is also particularly effective for mental and spiritual healings because conditioning, attachments and limiting beliefs are mental states which have emotional underpinnings.   Once unhealthy or suppressed emotions are released, this literally allows our light to shine more coherently. We can begin to see beyond duality consciousness. (See Bibliography for details of EFT and spiritual healing).

Another way to open the heart to love is to listen to music; music which is uplifting and inspiring; music that opens the heart chakra. This can be different for different people but it is something that resonates with a memory of a deep and loving connection, consciously or sub-consciously remembered. The practice of overtone chanting shows us that the sounds we hear contain resonating harmonics. This is why music is inspiring. We can gain access to these higher harmonics where there is greater coherence. "And harmonies unheard in sound create the harmonies we hear and wake the soul to the consciousness of beauty". Plotinus.

CHAPTER 26

ALL YOU NEED IS LOVE

'All you need is Love'
The Beetles

Music that opens the heart can fall into several categories. Sometimes it is the music itself which creates coherent heart rhythms, such as the classics; Mozart, Schubert; Bach, etc. Sometimes it is the words alongwith the music in modern day classics such as 'Amazing Grace', 'From a Distance' by Bette Midler, 'I Will Always Love You' by Dolly Parton, 'Heal The World' by Michael Jackson, 'My Heart Will Go On' by Celine Dion. These are just examples. It is important to find the music you most resonate with.

It's no coincidence that almost all songs are about love. It's also no coincidence that music has a profound effect on our feelings because everything is sound, music, resonance. Our physical voice and musical instruments are a lower aspect of the OM. When people are inspired to write songs, they are literally taking in spirit ('inspirare' in Latin means breath and spirit). Spirit is another word for spiralling light waves of consciousness which, as explained earlier, is what we are in essence. With inspiration we are really communicating with a higher aspect of consciousness, which is why most songs are about love and through them we get the message that love is the key. Remember we have already learned that breath and spirit 'inspirare' are Light. **So through Light we learn that Love is the key to becoming coherent with the Light.** Through Love we become coherent and attain non-locality. When we are non-local our consciousness is everywhere at once. This is **Realisation., Enlightenment, Omnipresence and ultimately Omnipotent. If you recall William Tiller's explanation of the potential inherent in coherent Light. The method of achieving that coherence is Love – it would have to be.**

Source breathed. The breath is the spiralling lightwaves, the sound of the breath is OM. OM is resonating frequencies throughout all dimensions and contains all information including the template of form, the morphic fields. OM creates vortex particles from light which creates matter which contains the strong and weak nuclear forces. The spinning of the vortex creating gravity and electromagnetism. Resonance is the re-sounding of wave frequencies. Waves are information carriers. Love creates coherent waves. Coherent waves are accumulated information leading to non-locality, (meaning everywhere at once, or omnipresent). The golden ratio spiral, being the universal fractal and the holder of coherence, is the ultimate morphic resonance field. It is the accumulation of all information existing everywhere at once. In other words it is OMnipresent (present everywhere), OMniscient (all knowing) and Omnipotent (all powerful).

Love enables OMnipresence, OMniscience and OMnipotence, where we see that all matter, including dark matter, is LIGHT. It is all LIGHT. We become One with the LIGHT through LOVE. Love really is, in the air. Love changes everything. Love is all around you. What the world needs now is Love, sweet Love. All you need is Love, because Love is coherent Light.

REFERENCES:

INTRODUCTION

CHAPTER 1 – THE BEGINNING IN A NUTSHELL

CHAPTER 2 – THE MISUNDERSTANDINGS
1. Aspden, H., 1994. "Experiments on Free Energy," *Nexus*, Feb.-March issue, page 50.
2. Brower, M., 1992. *Cool Energy*, MIT Press, Cambridge, MA, p.5
3. Eisen, J., editor, 1994. *Suppressed Inventions and Other Discoveries*, Auckland
4. Grotz, T., 1994. "Around the Free Energy World in Thirty Days," *New Science News*, volume III, number 2, pages 2-3.
5. O'Leary, B. 1989. *Exploring Inner and Outer Space*, North Atlantic Books, Berkeley, CA.
6. O'Leary, B. 1993a. *The Second Coming of Science*, ibid.
7. O'Leary, B. 1994. "Survival vs. Suppressed Science," *New Science News*, bolume III, number 2, page 1.
8. O'Leary, B. 1993b. "Sacred Science: Expanding the Box Even Further," *Proceedings of the International Forum on New Science*, edit. M. Albertson, IANS, Ft. Collins, Co.

9. Bearden, Thomas E., *The Excalibur Briefing*, Tesla Book Company, Greenville, Texas and Strawberry Hill Press, San Francisco (1988 and 1990).
10. King, Moray B., *Tapping the Zero-Point Energy*, Paraclete Publishing, P.O. Box 859, Provo, UT 84603 (1989).
11. Puthoff, Hal. 'Space propulsion: can empty space itself provide a solution? *Ad Astra*, 1997; 9(1): 42-6.
12. Puthoff, Hal. 'SETI: the velocity of light limitation and the Alcubierre warp drive: an integrating overview', *Physics Essays*, 1996; 9(1): 156-8.
13. Puthoff, Hal. 'Everything for nothing', *New Scientist*, 28 July 1990: 52-5
14. Puthoff, Hal. 'Far out ideas grounded in real physics', *Jane's Defense Weekly*, 26 July 2000; 34(4): 42-6.
15. Grotz, T., 1994. "Around the Free Energy World in Thirty Days," *New Science News,* volume III, number 2, pages 2-3.
16. www.unknowncountry.com

CHAPTER 3 – THE PHYSICS

1. Pribram, K.H., Brain and Perception: *Holonomy and Structure in Figural Processing* (Hillsdale, NJ: Lawrence Erlbaum, 1991).
2. In 1957, Smith, Purcell and Ramsey discovered the neutron has a slight electric dipole. From the book '*The New Science of the Spirit*' by David A. Ash, published by The College of Psychic Studies, London.
3. Boyer, T. 'Deviation of the blackbody radiation spectrum without quantum physics', *Physical Review*, 1969; 182: 1374.

CHAPTER 4 – THE DISHARMONY

CHAPTER 5 – HOLISTIC CONSCIOUSNESS
1. Pert, Candace, B. *'Molecules of Emotion'*, Simon and Schuster

CHAPTER 6 – A LITTLE HISTORY LESSON
1. White, Michael, *'Isaac Newton, The Last Sorcerer'*, Fourth Estate, London

CHAPTER 7 – IT'S ALL RELATIVE

CHAPTER 8 – A DIALECTIC THEORY

CHAPTER 9 – THE SOUL IS A HOLOGRAM

CHAPTER 10 – VORTEX RESONANCE
1. Callum Coats, *Living Energies*, Gateway Books, p7

CHAPTER 11 - CONSCIOUSNESS IS THE KEY
1. Greyson, Bruce, 'Near-Death Experiences and Spirituality', *Zygon 41* (2),
2. Reported in BBC2 Horizon, February 2003 and The Telegraph, London, 22 October 2000.
3. Reported in BBC2 Horizon, February 2003.
4. Radin D. and Nelson, R. 'Evidence for consciousness-related anomalies' *Behavioural and Brain Sciences*, 1987; 10: 600-1.
5. Jahn, R.G. et al., 'Correlations of random binary sequences with prestated operator intention: a review of a 12-year program', *Journal of Scientific Exploration*, 1997; 11: 350

CHAPTER 12 – THE BLASPHEMY, OR THE BEHAVIOUR OF LIGHT
1. Ehrenhaft, Prof. Felix, *Acta Physica Austriaca* (Vol. 4, 1950 and Vol. 5, 1951)
2. Quoted by Callum Coats *'Living Energies'* Gateway Books, p. 24.

CHAPTER 13 – MORPHIC RESONANCE OF LIGHT
1. Jenny, Hans, *'Cymatics'* Macromedia ISBN: 1-888-13807-6
2. Schiff, M. *The Memory of Water: Homeopathy and the Battle of New Ideas in the New Science* (Harpercollins, 1994): 26
3. Davenas, E., et al., 'Human basophil degranulation triggered by very dilute antiserum against IgE', *Nature*, 1988; 333(6176): 816-8.
4. Bastide, M. et al., 'Activity and chronopharmacology of very low doses of physiological immune inducers, *'Immunology Today*, 1985; 6: 234-5; L. Demangeat et al., Modifications des temps de relaxation RMN a 4MHz des protons du solvant dans les tres hautes dilutions salines de silice/lactose', *Journal of Medical Nuclear Biophysics*, 1992; 16: 135-45; B.J. Youbicier-Simo et al., 'Effects of embryonic bursectomy and in ivo administration of highly diluted bursin on an adrenocorticotropic and immune response to chikens', *International Journal of Immunotherapy*, 1993; IX: 169-80; P.C. Endler et al., The effect of highly diluted agitated thyroxin on the climbing activity of frogs', *Veterinary and Human Toxicology*, 1994; 36: 56-9.
5. Endler, P.C., et al., 'Transmission of hormone information by non-molecular means', *FASEB Journal*, 1994; 8: A400; F. Senekowitsch et al., 'Hormone effects by CD record/replay', *FASEB Journal*, 1995; 9: A392.

6. Benveniste, J. et al., 'Digital recording/transmission of the cholinergic signal', *FASEB Journal*, 1996, 10: A1479; Thomas, Y., et al, 'Direct transmission to cells of a molecular signal (phorbol myristate acetate, PMA) via an electronic device,' *FASEB Journal*, 1995; 9: A227; Aissa, J. et al., 'Molecular signalling at high dilution or by means of electronic circuitry', *Journal of Immunology*, 1993; 150: 146A; Aissa, J., 'Electronic transmission of the cholinergic signal', *FASEB Journal*, 1995; 9: A683: Thomas, Y. 'Modulation of human neutrophil activation by "electronic" phorbol myristate acetate (PMA)', *FASEB Journal*, 1996; 10: A1479.
7. Benveniste, J. 'Understanding digital biology', unpublished position paper, 14 June 1998; also interviews with J. Benveniste, October 1999.
8. See 6.
9. Benveniste, J., P. Jurgens et al. 'Transatlantic transfer of digitised antigen signal by telephone link', *Journal of Allergy and Clinical Immunology*, 1997; 99: S175.
10. Benveniste, J. et al. 'The molecular signal is not functioning in the absence of "informed" water', *FASEB Journal*, 1999; 13: A163.
11. Jibu, M., S. Hagan, S. Hammeroff et al., 'Quantum optical coherence in cytoskeletal microtubules: implication for brain function', *BioSystems*, 1994; 32: 95-209.
12. Del Guidice, E. and Preparata, G., 'Water as a free electric dipole laser', *Physical Review Letters*, 1988; 61:1085-88/
13. Laszlo, E. '*The Interconnected Universe*': Conceptual Foundations of Transdisciplinary Unified Theory (Singapore: World Scientific, 1995): 41.
14. Hameroff, S. '*Ultimate Computing*'; Jibu et al., 'Quantum optical coherence'

15. Del Guidice, E., et al., 'Electromagnetic field and spontaneous symmetry breaking in biological matter', *Nuclear Physics*, 1983; B275 (FS17): 185-99.
16. Hameroff, S. *Ultimate Comupting: Biomolecular Consciousness and Nano-technology* (Amsterdam: North Holland, 1987).
17. Jahn R., and B. Dunne, 'Science of the Subjective', *Journal of Scientific Exploration*, 1997; 11(2): 201-24.
18. Dunne, B.J., 'Co-operator experiments with an REG device', *PEAR Technical Note 91005*, December 1991.
19. Jahn, R. and Dunne, B., 'Margins of Reality: *The Role of Consciousness in the Physical World*' (London: Harcourt Brace Jovanovich, 1987).
20. Jahn, R. and Dunne, B., 'ArtREG: a random event experiment utilizing picture-preference feedback', *Journal of Scientific Exploration*, 2000; 14(3): 383-409.
21. Jahn, R.G. et al., 'Correlations of random binary sequences with prestated operator intention: a review of a 12-year program', *Journal of Scientific Exploration*, 1997; 11: 345-67.
22. Jahn, R., 'A modular model of mind/matter manifestations', *PEAR Technical Note* 2001.01, May 2001 (abstract).
23. Centre for Implosion Research, P.O. Box 38, Plymouth PL7 5YX. Tel: 01752 338438.

CHAPTER 14 – THE LAW OF RESONANCE

1. Gerber, R., *Vibrational Medicine* (Santa Fe: Bear and Company, 1988): 62.

CHAPTER 15 – PROOF OF SPEEDS FASTER THAN LIGHT

CHAPTER 16 – ELECTROMAGNETISM

CHAPTER 17 – GRAVITY AND THE STRONG AND WEAK NUCLEAR FORCES

CHAPTER 18 – DARK MATTER AND BLACK HOLES

CHAPTER 19 – SEEING WITH THE HEART

CHAPTER 20 – THE GOLDEN RATIO AND FRACTALS
1. Clarke, J. "SQUIDS", *Scientific American* (August 1994): pp46-53
2. Institute of HeartMath, 14700 West Park Avenue, Boulder Creek, CA 95006.

CHAPTER 21 – REINCARNATION
1. Stevenson, Ian, "Reincarnation: Field Studies and Theoretical Issues", *Handbook of Parapsychology*, edit. Benjamin B. Wolman, Van Nostrand Reinhold Company, New York (1977).
2. *Journal of the American Society for Psychical Research* 88: 207-219

CHAPTER 22 – THE HEART AND QUANTUM CONCEPTS
1. D.C. McClelland and J.B. Jemmott "Power Motivation, Stress and Physical Illness". *Journal of Human Stress*, Vol. 6 (1980) pp 6-15
2. D.C. McClelland and C. Kirshnit "The Effect of Motivational Arousal Through Films on Salivary Immunoglobulin A" *Psychology and Health,* Vol. 2 (1988) pp 31-52.
3. Studies on affiliative trust are summarised in D.C. McClelland "Motivational Factors in Health and Disease", *American Psychologist*, Vol. 44 (1989) pp 675-683.
4. 15A J. Motz "Everyone an Energy Healer: The TREAT V Conference in Santa Fe" Advances Vol. 9 (1993): pp 95-98

5. 'Evidence of a Primary Perception in Plant Life', *International Journal of Parapsychology*, vol. 10, no. 4, Winter 1968, pp. 329-48.
6. 15B R.C. Byrd "Positive Therapeutic Effects of Intercessory Prayer in a Coronary Care Unit Population" *Southern Medical Journal*, Vol. 81, (1988) pp 826-829

CHAPTER 23 – AURAS AND RADIATION

CHAPTER 24 – THE CAUSE OF CANCER

1. Coghill Research Labs, www.cogreslab.co.uk, Powerwatch website, www.powerwatch.org.uk, Electromagnetic Hazard & Therapy www.em-hazard-therapy.com, Mobile phone Health Research Group www.mthr.org.uk, SITEL – a Belgian association of electronic engineering professionals and companies www.sitel.org This is a sample of websites providing a list of references to published scientific papers.
2. The School and Clinic of Electro-Crystal Therapy, 117 Long Drive, South Ruislip, Middlesex HA4 OH1
3. Popp, F.A., 'MO-Rechnungen an 3.4-Benzpyren und 1,2-Benzpyren legen ein Modell zur Deutung der chemischen Karzinogenese nahe', Zeitschrift fur Natur-forschung, 1972; 27b: 731; Popp, F.A., 'Einige Moglichkeiten fur Biosignale zur Steurung des Zellwachstums', *Archiv fur Geschwulstforschung*, 1974; 44: 295-306.
4. Ruth, B. and Popp, F.A., 'Emperimentelle Untersuchungen zur ultraschwachen Photonemission biologisher Systeme', *Zeitschrift fur Naturforschung*, 1976; 31c: 741-5.
5. Rattemeyer, M., Popp, F.A. and Nagl, W., *Naturwissenschaften*, 1981; 11: 572-3.
6. Popp, F.A. and Chang Jiin-Ju, 'Mechanism of interaction between electromagnetic fields and living systems', *Science in China* (Series C), 2000: 43: 507-18.

7. Nobili, R. 'Schrodinger wave holography in brain cortex', *Physical Review A*, 1985; 32: 3618-26. Nobili, R. 'Ionic waves in animal tissues', *Physical Review A*, 1987; 35: 1901-22.
8. Frohlich, H. 'Long-range coherence and energy storage in biological system', *International Journal of Quantum Chemistry*, 1968; 2: 641-9.
9. Frohlich, H. 'Evidence for Bose condensation-like excitation of coherent modes in biological systems', *Physics Letters*, 1975, 51A: 21; see also Zohar, D., *The Quantum Self* (London: Flamingo, 1991): 65.

CHAPTER 25 – MUSIC OF THE SPHERES

1. Tompkins, P. and Bird, C. 1973 *The Secret Life of Plants*, Harper & Row, Inc.
2. Campbell, D. 1997 *The Mozart Effect*, Hodder & Stoughton, London.

CHAPTER 26 – ALL YOU NEED IS LOVE

BIBLIOGRAPHY, RECOMMENDED READING AND WEBSITES

Alexandersson, Olof, *Living Water*, Gateway (Gill & Macmillan Ltd., Dublin) (1976)

Ash, David A. PhD., *The New Science of the Spirit*, The College of Psychic Studies, London (1995)

Bailey, Alice A. *The Consciousness of the Atom*, Lucis Press Ltd. (1922)
Bailey, Alice A. *A Treatise on Cosmic Fire*, Lucis Press Ltd., (1925)
Bailey, Alice A. *From Intellect to Intuition*, Lucis Press Ltd., (1932)

Blavatsky, H.P., *The Secret Doctrine*, Quest Books, The Theosophical Publishing House, Abridgement published in 1966. First abridgement published in 1907.

Brennan, Barbara Ann, *Hands of Light*, Bantam Books (1988)

Burr, Harold Saxton, *Blueprint for Immortality: Electric Patterns of Life Discovered in Scientific Breakthrough*, C.W. Daniel Co. Ltd.,

Campbell, Don, *The Mozart Effect*, Hodder & Stoughton, (1997)

Cayce, Edgar, *A Seer Out of Season*, The Aquarian Press (1989) (and all the works of Edgar Cayce)

Coats, Callum, *Living Energies*, Gateway Books, Bath, UK, (1996)

Cockell, Jenny, *Across Time and Death*

Courteney, Hazel   *Evidence for the Sixth Sense*, Cico Books, London (1999)

Cowan, David, and Rodney Girdlestone, *Safe As Houses*, Gateway Books, Bath, U.K. (1996)

Darwin, Charles, *The Origin of Species*, First published by John Murray, (1859)

Davidson, John, *The Secret of the Creative Vacuum*, Saffron Walden, the C.W. Daniel Company Ltd. (1989)

Emoto, Dr. Masaru, *Messages from Water*, HADO Kyoikusha Co. Ltd. (1999-2002)

Garfield, Laeh Maggie, *Sound Medicine*, Celestial Arts, California (1987)

Gerber, Richard, M.D., *Vibrational Medicine*, Bear & Company, New Mexico (1988)

Hapgood, Charles H., *Maps of the Ancient Sea Kings*, Adventures Unlimited Press, USA (1996)

Hartmann, Thom, *The Last Hours of Ancient Sunlight*, Mythical Research Inc. (1998, 1999) and Hodder & Stoughton, UK (2001)

Jeans, Sir James, *The Universe Around Us*, Cambridge University Press, (1945)

Jenny, Dr. Hans, *Cymatics*, Macromedia, Newmarket, USA, Vol 1. orig. published in 1967, Vol. 2 orig. published in 1974. First printing of combined volumes 2001.

Jung, Carl G., *Man and his Symbols*, Picador (1964)

Krishnamurti, J., *The Wholeness of Life*, Victor Gollancz Ltd. (1991) (and all the works of J. Krishnamurti)

Livio, Mario, *The Golden Ratio*, published by Headline Book Publishing.

Martineau, John, *A Little Book of Coincidence*, Wooden Books

McTaggart, Lynne, *The Field*, HarperCollins Publishers, London (2001)

Mills, Dr. Antonia, *Amerindian Rebirth: Reincarnation Belief Among North American Indians and Inuit* (University of Toronto Press)

Moody, Dr. Raymond, *Life After Life*, Books (1976)

Motoyama, Hiroshi, *Theories of The Chakras: Bridge to Higher Consciousness.*, The Theosophical Publishing House, Illinois, USA, Madras, India, London, England (1981)

O'Donohue, John, *Divine Beauty*

O'Leary, Brian, PhD., *Exploring Inner and Outer Space*, North Atlantic Books, California (1989)
O'Leary, Brian, PhD., *The Second Coming of Science*, North Atlantic Books, California (1993)
O'Leary, Brian, PhD., *Miracle In The Void*, Kamapua'a Press, Hawaii (1996)
O'Leary, Brian, PhD., *Reinheriting The Earth*,

Paulsen, Norman, *The Christ Consciousness*, The Builders Publishing Co., Utah, USA (1980, 1984, 1994)

Pearsall, Paul, PhD., *The Heart's Code*

Pert, Candace, PhD., *Molecules of Emotion*, Scribner, USA (1997) and Simon & Schuster, UK (1998)

Siegel, Bernie, M.D., *Love, Medicine and Miracles*, Arrow (1988)

Tiller, William, Prof., *Conscious Acts of Creation*, Paviour Pub. (2001)
Tiller, William, Prof., *Science & Human Transformation: Subtle Energies, Intentionality and Consciousness*, Pavior Pub. (1997)

Tompkins, Peter and Bird, Christopher, *The Secret Life Of Plants*, Harper & Row (1973)
Tompkins, Peter and Bird, Christopher, *Secrets of the Soil*, Harper & Row (1989)

Wall, Vicky, *The Miracle of Colour Healing*, Aquarian/Thorsons (1990)

White, Michael, *Isaac Newton, The Last Sorcerer*, Fourth Estate, London (1997)

Winter, Dan, PhD., *Alphabet of the Heart*

EMOTIONAL FREEDOM TECHNIQUES www.emofree.com
NATIONAL FEDERATION OF SPIRITUAL HEALERS www.nfsh.org.uk
SPIRITUAL EMERGENCE /AWAKENING EXPERIENCES www.holotropic.com
HEALING/MEDITATION/SPIRITUAL RETREAT www.yogaville.org & www.noetic.org
KRISHNAMURTI PHILOSOPHY/RETREAT www.kfoundation.organd www.krishnamurticentre.org.uk
KIRLIAN PHOTOGRAPHY www.electrocrystal.com
INSTITUTE OF NOETIC SCIENCE www.noetic.org
NEW ENERGY MOVEMENT www.newenergymovement.org
ZERO POINT TECHNOLOGY www.disclosureproject.org

Lightning Source UK Ltd.
Milton Keynes UK
25 October 2010

161843UK00011B/120/P